BLUEPRINT FOR A BATTLESTAR

BLUEPRINT FOR A BATTLESTAR

SERIOUS SCIENTIFIC EXPLANATIONS BEHIND SCI-FI'S GREATEST INVENTIONS

ROD PYLE

New York

STERLING
New York

An Imprint of Sterling Publishing Co., Inc.
1166 Avenue of the Americas
New York, NY 10036

First published in the UK in 2016

ISBN 978-1-4549-2134-9

For information about custom editions, special sales, and premium and corporate purchases, please contact Sterling Special Sales at 800-805-5489 or specialsales@sterlingpublishing.com.

Manufactured in China

www.sterlingpublishing.com

Design by Aurum Press

CONTENTS

INTRODUCTION

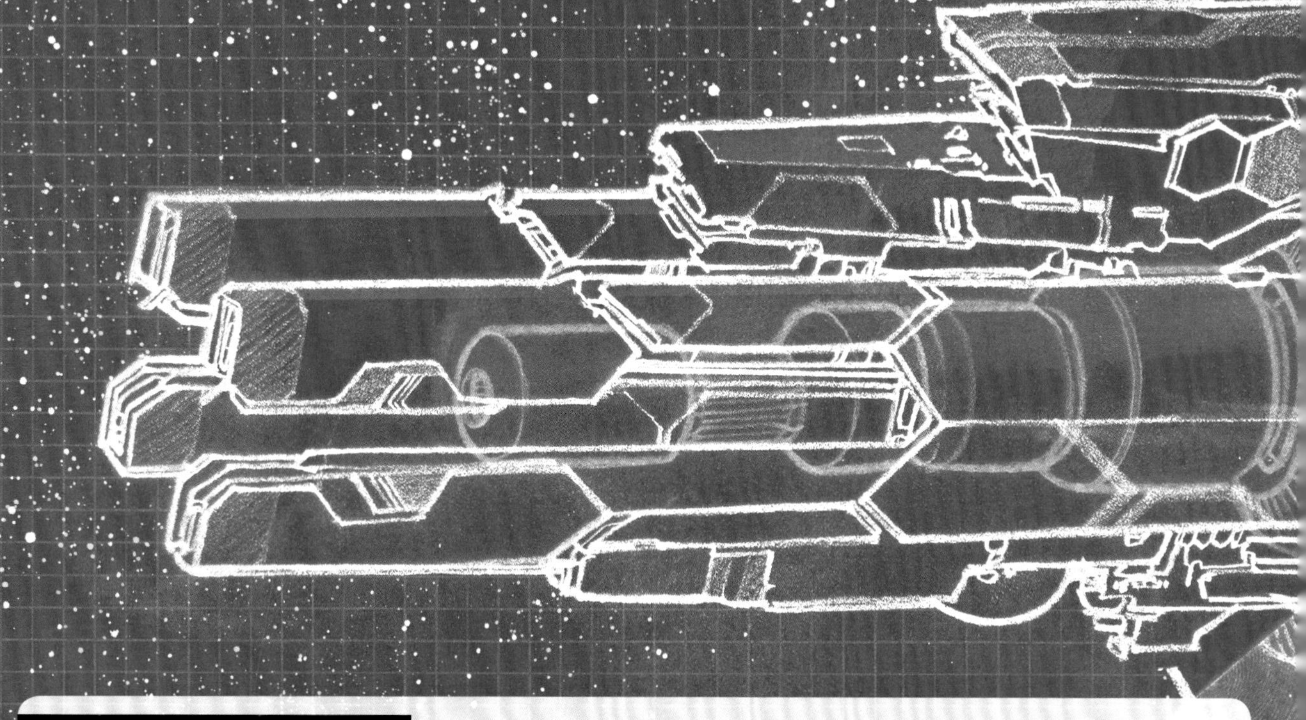

DESIGNING THE FUTURE

The cool thing about the future is that it always gives you something to look forward to.

It's uncouth to quote yourself, so we'll just leave the above comment uncredited...but it's true. Writing this book has been by turns great fun and vastly challenging, but the overall picture—though not what was foreseen in my childhood at all—is one of a potentially bright and shining tomorrow.

Teasing cool and interesting topics out of contemporary science and tech journalism, as well as seeking themes in quality sci-fi, is tricky at the best of times. Movie directors often miss the mark, science fiction writers can have agendas of their own, and every creator is subject to market forces that tend to proclaim the importance of whatever boosts the bottom line. One must paw through vast amounts of material for the best and most common themes—the cream of future visions—in order to rise to the top.

In the past decade of writing books and science journalism, I've been exposed to a lot of great thinkers—some rightfully famous, others less so but equally deserving of it. The process has been inspiring, engaging, and humbling. Working at Caltech and various NASA centers, I would find myself in rooms full of hyper-intelligent (and mostly young) minds, and wonder how these young people became so brilliant in such a short period of time. Interviewing grad students at Caltech, Stanford, UCLA, and MIT left a similar impression—these are the much maligned Millennials, the cannon-fodder for so much pop-journalism. My overriding impression was of bright, energetic, and impassioned youth who are excited to be dreaming

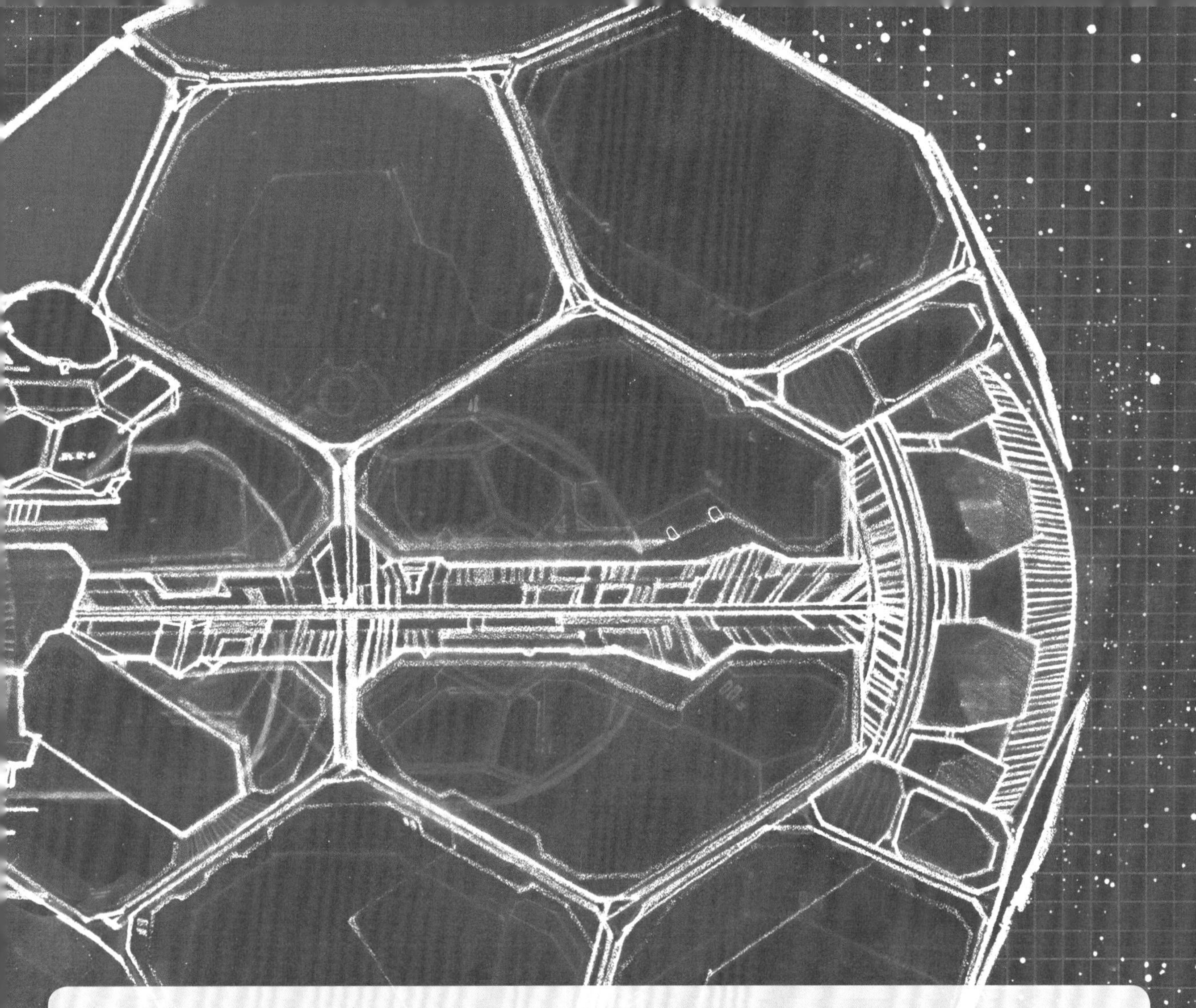

of, and designing, the future. There will be bumps and challenges for them, as both legal and ethical boundaries lag far behind invention, but I'm convinced they will figure it out.

A lot of research, some sleuthing, and a good bit of intuiting goes into a book such as this. I'm fortunate to have spent two decades writing about science, technology, and spaceflight—current and future trends, as well as copious amounts of history. An awareness of history helps to inform one's thoughts about the future, so it's important. Add to this some great input from smart science and technology practitioners, vast amounts of wonderful resources (all of us should have ready access to the fee-based archives that universities and government do) and a healthy blend of cautious optimism, and you've got *Blueprint for a Battlestar*.

All that said, there will be both errors in reporting and prediction (the former is unfortunate, the latter inevitable). My crystal ball is cloudy. My Magic 8-Ball said 'Reply hazy, try again' more than once. More to the point, even primary references occasionally disagree. Nonetheless, any mistakes are mine, so feel free to email with observations of factual errors. But with regard to mistaken predictions, likewise feel free enjoy a quiet, warm inner glow of being right. If anything, I hope that I have underestimated humanity's genius and intrinsic goodness.

I hope you enjoy the book and these visions of where we are and what's next. Read on.

WEAPONS OF THE FUTURE

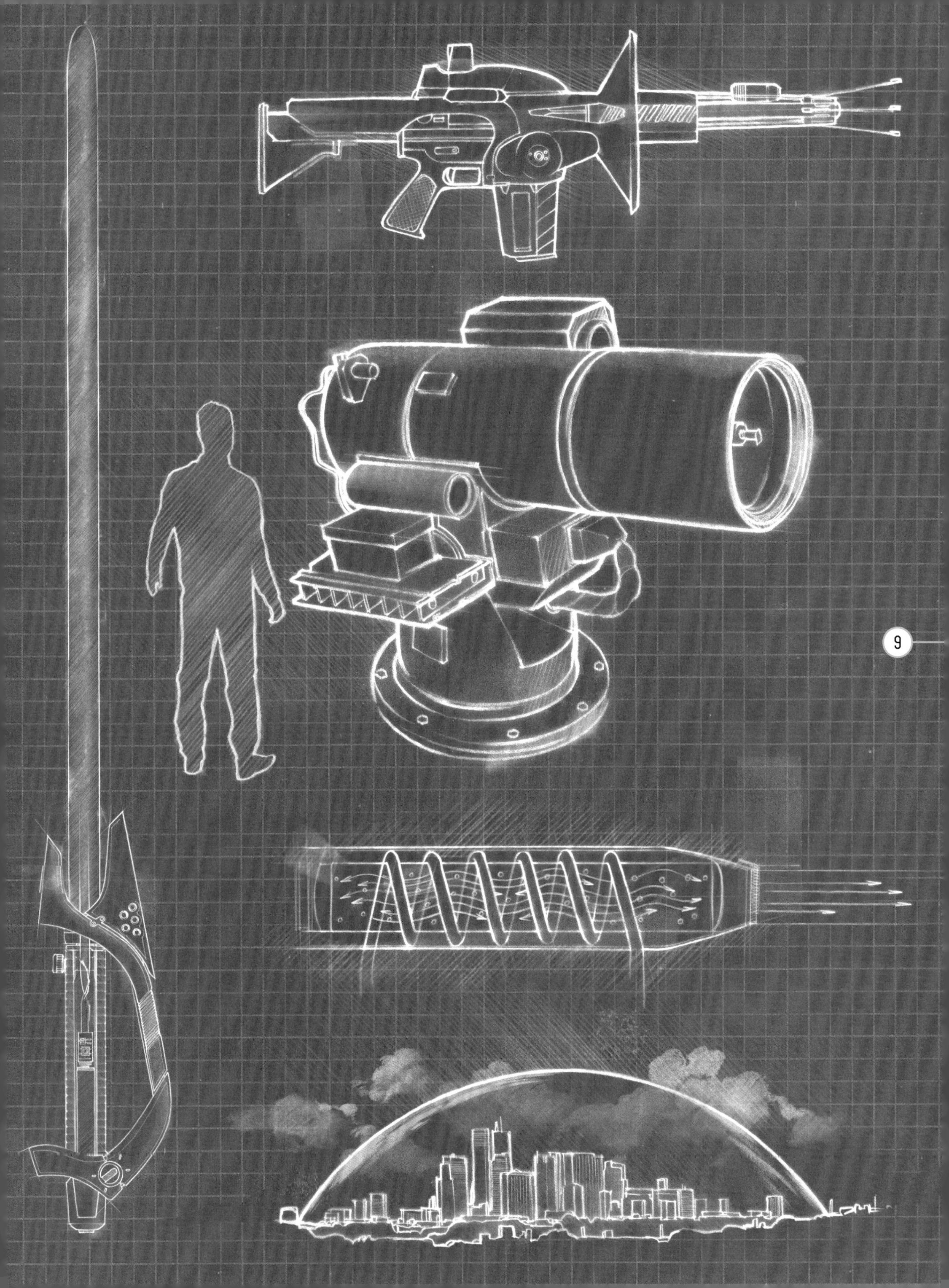
9

DEATH FROM ABOVE: BUILDING A DEATH PLANET

Without doubt, the technological celebrity of *Star Wars Episode IV: A New Hope* was the death Planet known as the Death Star. It was the threat that hovered above a world...the target of the heroic pilots in their X-Wing fighters and the home base that Darth Vader so tenaciously defended with his wicked Twin Ion Engine (TIE) Fighters.

TACTICAL CONCERNS

The death planet seemed so effective that it was even the subject of a public petition sent to the White House in 2012. The suggestion was that the US should build its own Death Star, presumably to maintain law and order on some planet other than Earth. The tongue-in-cheek document garnered more than 25,000 signatures, enough to warrant an equally silly response from the Obama administration. They said the cost of such a device was estimated at about $850 quadrillion (which is even larger than the US debt,) and that it would take 833,000 years to create enough metal to fabricate it. The response further elaborated that the US government did not support blowing up planets and had tactical concerns about a weapon which could be destroyed by a tiny, battered one-man fighter.

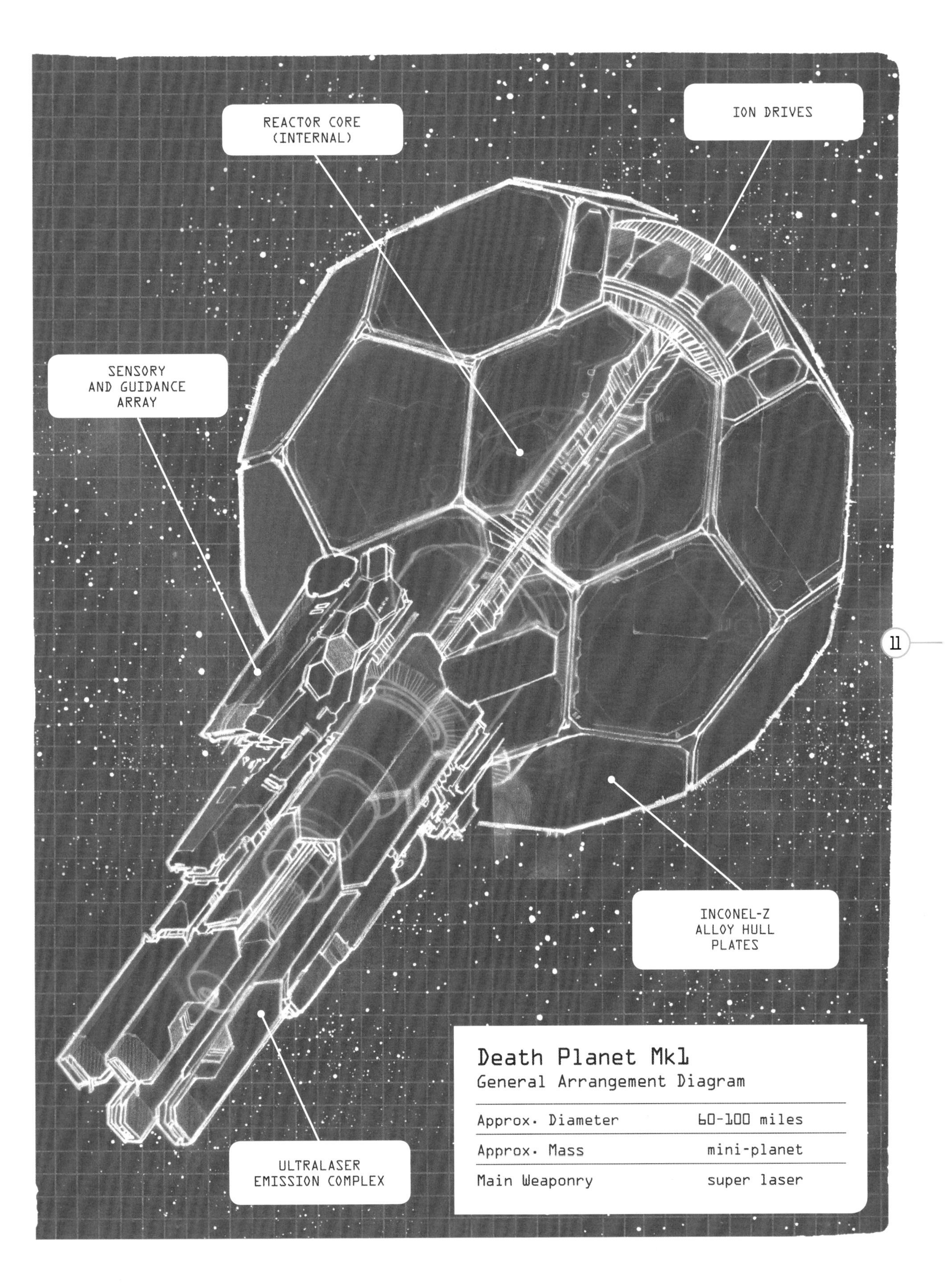
REACTOR CORE
(INTERNAL)
ION DRIVES
SENSORY
AND GUIDANCE
ARRAY
INCONEL-Z
ALLOY HULL
PLATES
ULTRALASER
EMISSION COMPLEX
Death Planet Mk1
General Arrangement Diagram
Approx. Diameter 60-100 miles
Approx. Mass mini-planet
Main Weaponry super laser

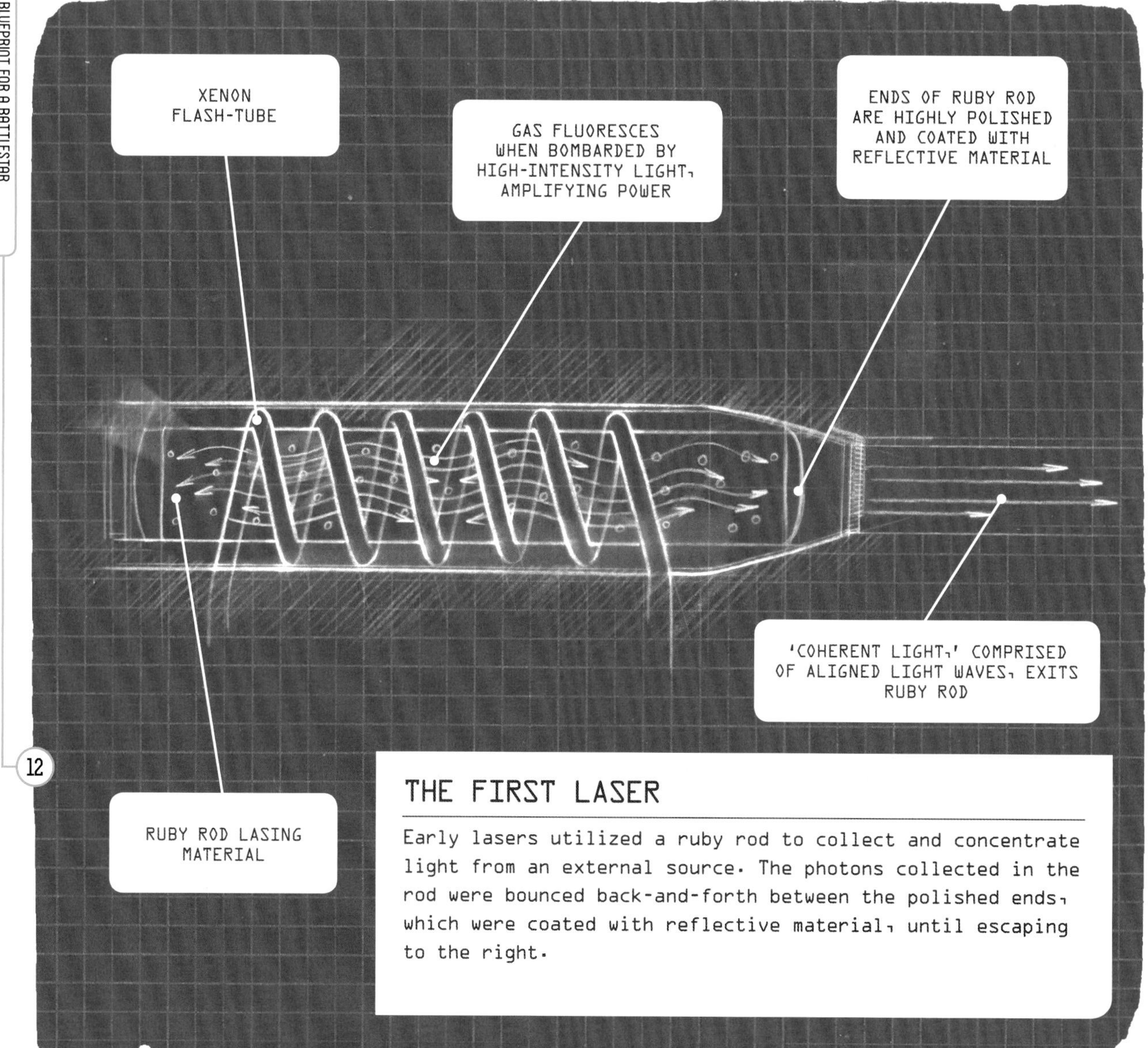

THE FIRST LASER

Early lasers utilized a ruby rod to collect and concentrate light from an external source. The photons collected in the rod were bounced back-and-forth between the polished ends, which were coated with reflective material, until escaping to the right.

NASA may not be working on a death planet. But what would it take to actually build one? The first problem to overcome is sheer size: The death planet, as represented in the *Star Wars* films, is really big. Various sources have estimated its diameter at somewhere between 60–100 miles. At this size it would put most of the inhabitants of the asteroid belt to shame and, even though basically hollow, would have a small gravitational field and essentially be a mini-planet, if you will.

If something that size was placed in Low Earth Orbit—the region around our planet extending to an altitude of 60–1,200 miles—a goodly chunk of it would be hanging down into the atmosphere, dragging and heating up and rapidly reentering. It would most likely crash into the ocean.

Just getting the death planet into orbit in the first place wouldn't be easy. After spending nearly a million years to fabricate the metal, it would take an untold number of launches to get all that mass up there...enough to permanently poison the atmosphere.

There is one way around this particular problem, however. There is plenty of metallic ore in the asteroids that roam the solar system. Most of it is in the asteroid belt and the Kuiper Belt (outside of

Pluto's orbit,) but there are also plenty of rogue wanderers. Grabbing a number of large ones and smelting their ores down would net many millions of tons of metal that is already up in space...no bulk launches necessary for building materials. Of course, powering the mining operation itself would require solar panels, possibly thousands of miles across.

If built on Earth getting the death planet into space would be daunting. The largest flying machine ever created was the Apollo program's Saturn V rocket, which weighed over five million pounds, when its fuel is included. It was about the size and mass of a World War II navy destroyer and, had it exploded, would have had the power of a small atomic bomb. All this explosive power was required simply to propel the tiny 12,000-lb capsule at its nose to the moon and back. Moving the death planet would require something hugely more powerful.

Supposing a death planet could actually be built and powered, it would still lack the very nasty weapon the original version had. It is never explicitly said exactly what kind of weapon the death planet used to destroy Alderaan, but some documentation refers to the weapon as a "Super Laser." Laser is an acronym for Light Amplification by Stimulated Emission of Radiation (LASER.) This is a fancy term for 'coherent' light, a form of light in which all the waves cooperate to move essentially in unison. This can create a powerful beam that carries a lot of energy a long way, if the laser is powerful enough. Lasers have been around for about 50 years and are well understood devices, currently in use in everything from DVD players to military weapons.

Today's most powerful lasers don't even use electricity to power them. Instead they employ gases that are forced explosively through a large tube at high velocities. In the correct configuration, a gas that changes temperature rapidly enough can emit light, in this case coherent light. But you need *a lot* of gas, moving *very* quickly, to do this. The upside is that a lot of very powerful, hot light can be created for a brief time. The largest such lasers known today are the MIRACL (Mid-Infrared Advanced Chemical Laser,) made by the US Navy, and a US-Israeli collaboration called the Tactical High Energy Laser (THEL.) Lasers in this size and power range, the current upper limit, can shoot down artillery shells and small battlefield rockets up to five feet long and well under a foot in diameter. That's a bit smaller than Alderaan.

That planet was around 7,700 miles in diameter, or about the same size as Earth (just shy of 8,000 miles.) To destroy it would take the equivalent of over a sextillion (or a billion trillion) artillery shells. The 'super laser' would need to be enormous. The *Star Wars* version also vaporizes the planet in just two to three seconds, far quicker than today's battlefield lasers can destroy those artillery shells. So for a gas laser, you would need a small planet-full of gas (bigger than the death planet itself) to fire up the laser—or all the power that could be made by the combined generating stations of Earth in a few trillion years (for comparison, the entire universe is just over 12 billion years old).

Elsewhere in the official *Star Wars* universe, it is mentioned that the weapon is actually powered by 'hypermatter.' Regardless of what powers it, vaporizing matter gives off energy. How much matter would need to be flashed out of existence to create that much power? About the equivalent of the mass in Mount Everest is how much. We do have a lot of mass on Earth, enough to make many thousands of Mount Everests so as long as we're willing to dig enormous holes in our planet. The problem is to figure out a way to convert it into energy.

Bear in mind that, using this system of reference, one gram of matter contains as much energy as 21.5 kilotons of TNT high explosive, or about the same explosive force of the plutonium atom bomb dropped on Nagasaki in the Second World War. So how many grams are in Mount Everest? The mountain contains about 365 cubic miles of mass, or about 6,399,000,000,000 metric tons. That's a huge amount of explosive energy to be released if we vaporized it, making it possibly more practical than a gas-powered death planet.

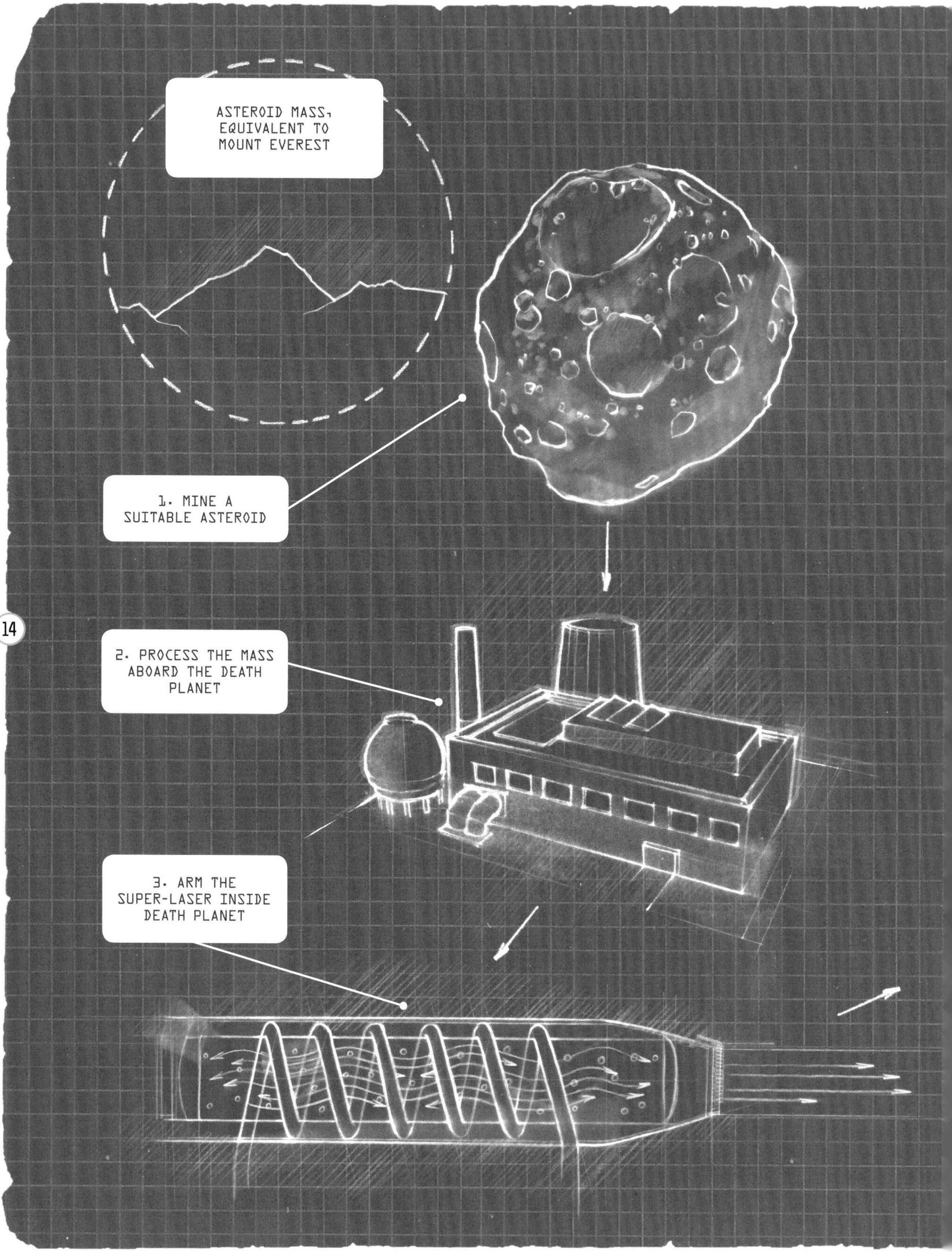

ASTEROID MASS,
EQUIVALENT TO
MOUNT EVEREST

1. MINE A
SUITABLE ASTEROID

2. PROCESS THE MASS
ABOARD THE DEATH
PLANET

3. ARM THE
SUPER-LASER INSIDE
DEATH PLANET

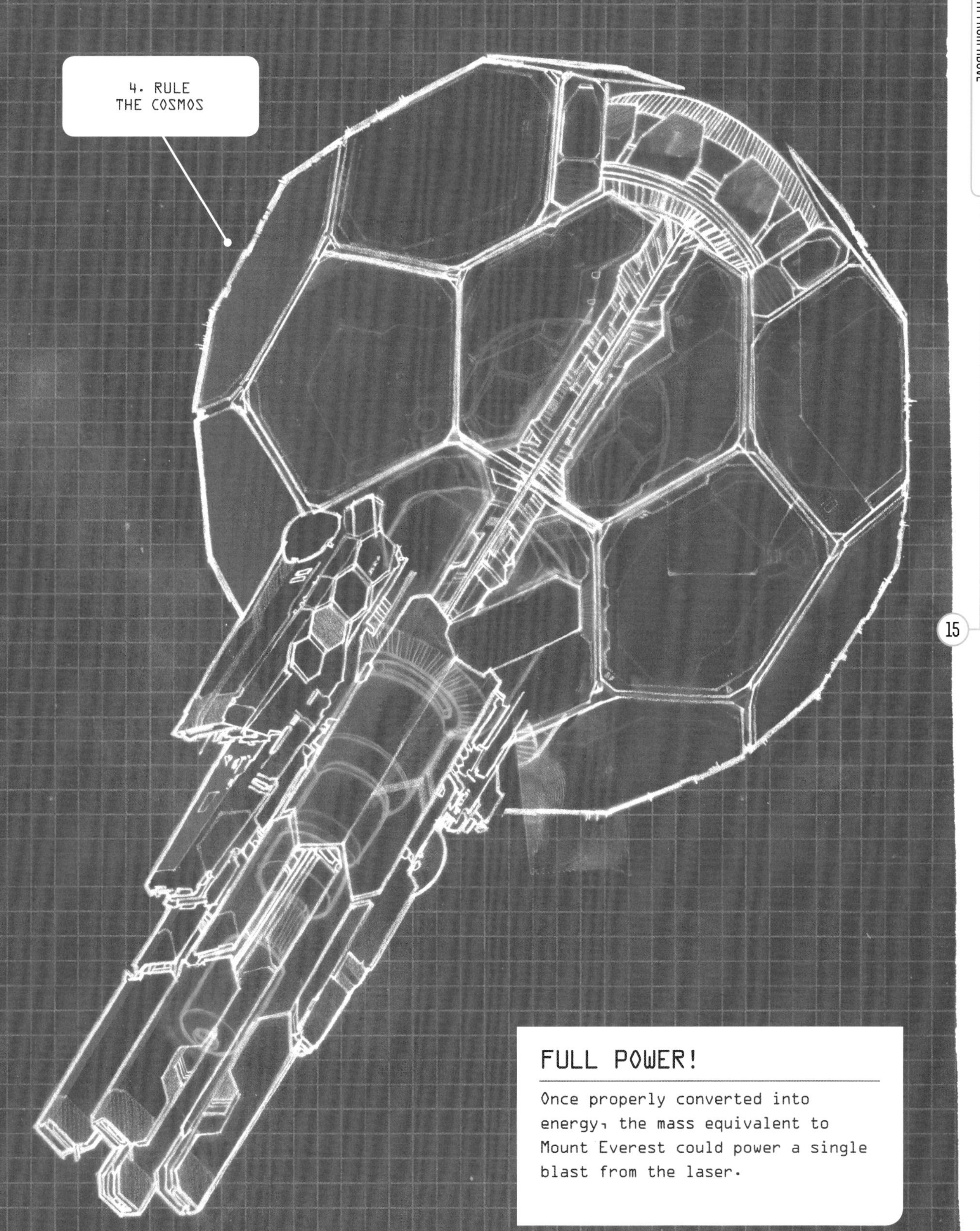

FULL POWER!

Once properly converted into energy, the mass equivalent to Mount Everest could power a single blast from the laser.

THE ULTIMATE WEAPON: MAKING A DEATH RAY

Death rays have been a staple of science fiction for decades, but people have fantasized about death-dealing beams of destruction for much longer than that. Since humans first went to war, the idea of an omnipotent weapon of destruction has been with us, made partially obsolete only by the atom bomb.

AN ANCIENT WEAPON

Archimedes, that great astronomer, inventor, mathematician, engineer, and all-around troublemaker of ancient Greece, was the first to go on record with the idea. During the Siege of Syracuse, between 214 and 212 BCE, he was said to have designed a "heat ray" that could sink the ships of the invading Roman fleet. Using a series of mirrors to focus the rays of the Sun to a high temperature, it caused the combustible materials, such as wood and tar, aboard to catch fire.

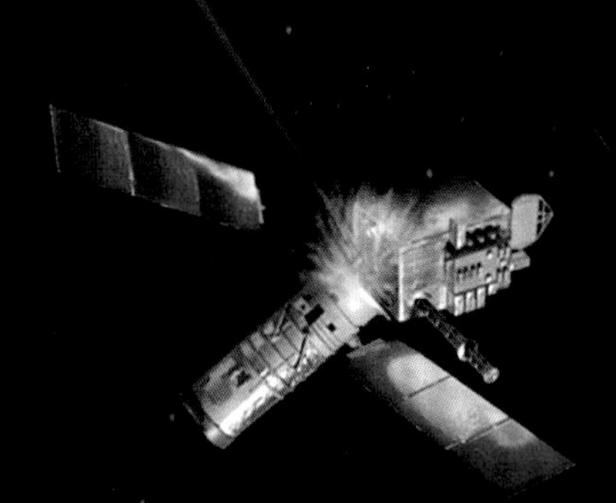

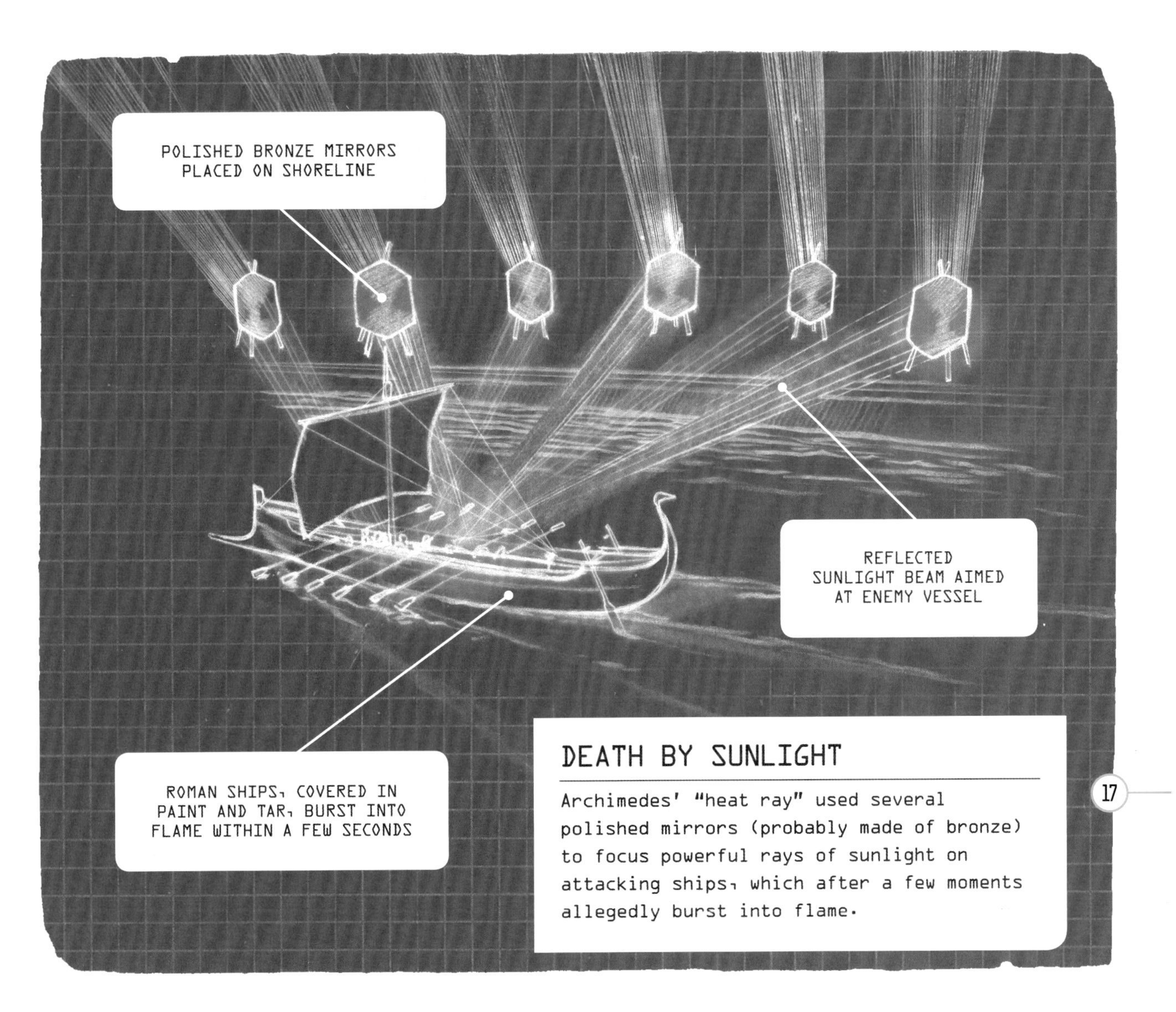

DEATH BY SUNLIGHT

Archimedes' "heat ray" used several polished mirrors (probably made of bronze) to focus powerful rays of sunlight on attacking ships, which after a few moments allegedly burst into flame.

Archimedes died when the Romans finally took Syracuse, and the knowledge of his 'death ray' perished with him. Attempts to recreate the event have never quite succeeded. Cloud cover, water vapor in the air, the motion of the target, poorly trained soldiers operating it—all of these could cause the ray to fail. But in principle, on a clear day, given enough time, it *could* work.

The next appearance of death ray—in the form of a heat ray variant—showed up in *The War of the Worlds*, British author H.G. Wells' 1897 tale of a Martian attack on Earth. The invading Martians have parabolic dishes that burn everything in their path, which completely sweep aside the British army and other defending Earth forces. It wasn't clouds or water vapor that knocked out the Martians in the end, but the common cold.

In the 20th century another form of death ray was devised by no less a genius than Nikola Tesla, famed for the invention of alternating current. Tesla published his ideas in the *New York Sun* in 1934, among other venues. He called it a 'peace ray' or 'teleforce,' but the much less the much less pacifist sounding 'death ray' was the name that stuck.

The device, apparently never built, used energy from a huge Van de Graff generator tower, a machine that creates copious amounts of static electricity (a much magnified version of what happens when you shuffle your wool-stockinged feet across a carpet on a dry day.) Small metallic particles would be hurled

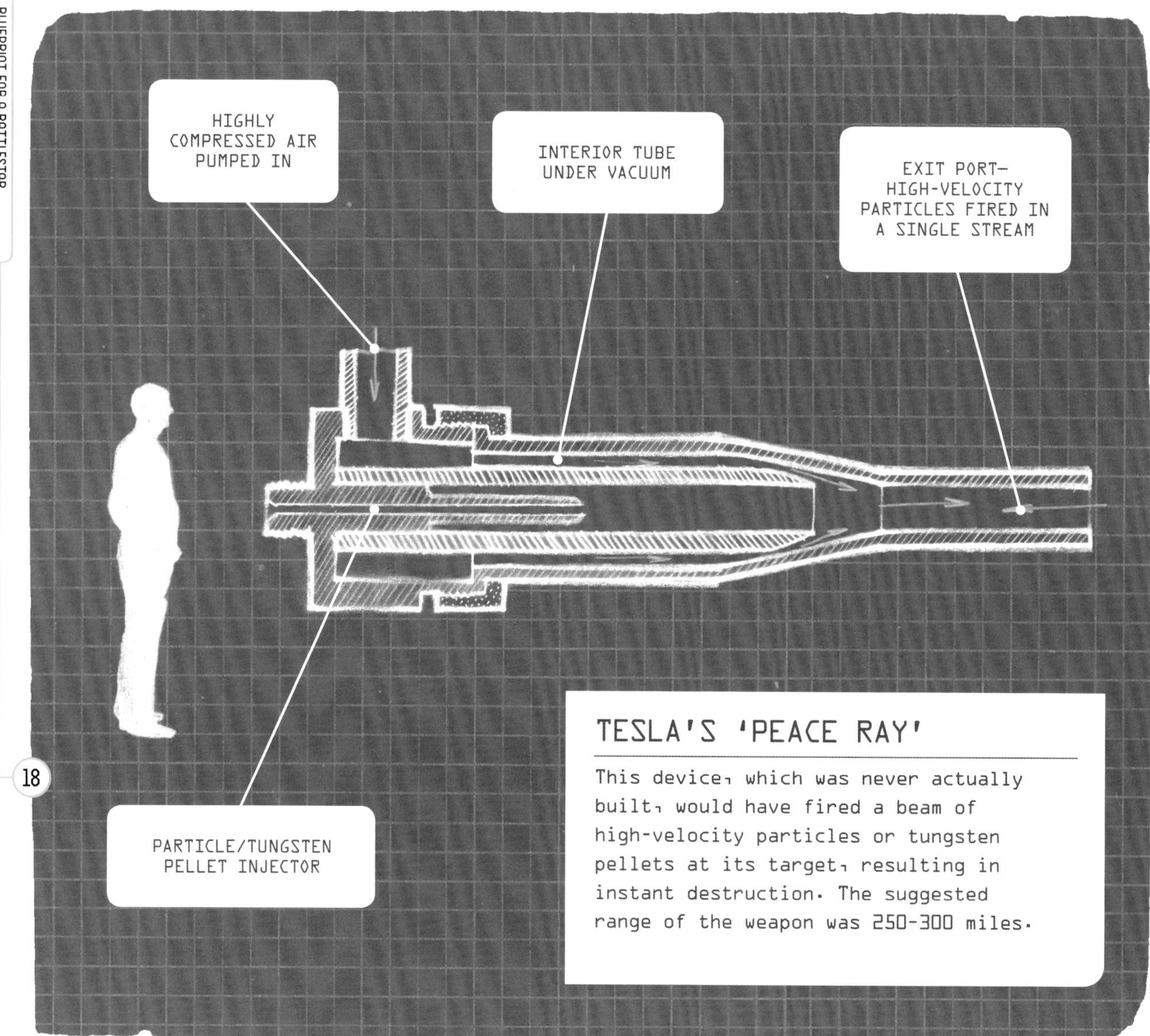

TESLA'S 'PEACE RAY'

This device, which was never actually built, would have fired a beam of high-velocity particles or tungsten pellets at its target, resulting in instant destruction. The suggested range of the weapon was 250-300 miles.

at extremely high velocities out of an open-ended tube (which had a vacuum inside.) Tesla claimed these tiny pellets would travel at about 48 times the speed of sound via electrostatic repulsion. It would take 60 million volts to power the peace ray, which would have to be generated by a huge generating station of Tesla's design.

Tesla claimed that his ray would have a range of hundreds of miles and would bring down any aerial machine that dared to cross its path. But he failed to secure funding and his 'peace ray' was never tested.

During World War II, the search for a war-winning secret weapon led the Germans to look at a city-burning heat ray as a means to deliver a knockout blow to the Allies. Just like the Martian version, this beam of death would also come from outer space. Its mastermind was the German rocketeer Hermann Oberth, who in the 1920s came up with the idea of building a giant concave mirror in orbit. His claims that it was intended for peaceful purposes such as 'illuminating ports' and 'thawing rivers' rang hollow with the coming to power of the Nazis in the 1930s, who were much more interested in its potential for burning great swaths of doom through enemy territory. At least one version of the weapon was to be 300 miles wide, with a full-time crew aboard, who would breath air generated, oddly enough, by an orbital pumpkin patch.

The idea was quietly forgotten for decades, until US President Ronald Reagan brought them back with a vengeance in the 1980s. The former Hollywood actor first mentioned his audacious plan just over a year after taking office, and by 1984 the US military was getting serious about the Strategic Defense Initiative (SDI,) in which death rays were to play a large part, this time to shoot down nuclear missiles.

Many in the technical fields thought that the goals —to render the United States safe from a Soviet nuclear missile strike—were unrealistic and the technologies required were unattainable. But defense contractors pushed ahead in the hope of big government contracts and the press quickly began referring to the overall program as '*Star Wars*.'

While the US and Soviet Union had both worked on defensive missile systems for years, they did not meet with much success. Firing counter-missiles to take down incoming enemy ballistic missiles was risky. It would take just one MIRV (a multiple independently targetable re-entry vehicle, essentially a single missile warhead that can split into 12 separate thermonuclear weapons) getting past the missile shield to cause massive destruction.

So SDI sought a better plan...and what could be better than a missile-wrecking heat ray? The idea was to hit enemy missiles before re-entry (the final descent to the target.) So, naturally, the '*Star Wars*' defense shield would be best deployed in space. The question was: could massive defense lasers actually be placed in orbit?

Lasers capable of tracking and hitting an enemy missile as it climbed into the sky would be complex and heavy. Many would be needed, and they would have to be in an orbital position close enough to the Soviet launch sites to be effective—a proximity which would also make them vulnerable to Soviet anti-satellite weapons. So these space-based laser would need their own proximity defense weapons. Most of these problems were deemed solvable (at least by the optimists,) but the massive power required was the toughest challenge. Batteries? Nuclear generators? Massive capacitors? Chemical fuels? Everything was considered.

One solution was to revive an ancient idea and place giant mirrors in orbit. Ground-based lasers would be bounced off these on to enemy targets. A precursor to such a system was actually tested on a Space Shuttle flight in 1985 with some success. The one big drawback was that weather conditions between the ground-based laser and the satellite could seriously compromise the defensive weapon's effectiveness.

Another approach experimented with by the SDI was a particle beam (not unlike Tesla's peace ray,) called a neutral particle beam. This device is similar to any other particle-beam generator, except that the particles emitted are neutralized (by adding or subtracting electrons) to keep the beam from diverging after it leaves the 'gun.' If not neutralized, the particles would tend to repel each other. An experiment with a low-power version of this weapon was conducted in space by firing a rocket with the beam-emitter and was successful, but so far as anyone knows, no such weapon has since been deployed. Such weapons are theoretically capable of generating bursts in the gigajoule range, equivalent at the point-of-impact to about 500 pounds of TNT... certainly enough to disable even hardened ICBMs.

Even more devastating may be the chemical laser, experimented with both as part of SDI and afterward. These lasers use a number of different chemical reactions to accomplish the creation of a bright pulse of coherent light. A fuel such as chlorine is mixed with hydrogen peroxide. The reaction of this mixture creates a mass of highly energetic oxygen molecules. Nitrogen is then forced into the system under high pressure, pushing the oxygen molecules through a mist of iodine. The energy of the oxygen molecules is transferred to the iodine, resulting in a very powerful burst of light. This light is bounced between a set of mirrors until it reaches sufficient intensity to escape. It is directed toward its target with mirrors and navigational aiming systems (in space, these could be gyroscopes and maneuvering thrusters.) The brief but incredibly intense beam is directed at the enemy missile, and melts enough of it to cause it to fail.

A similar system, called a deuterium fluoride laser, was tested in 1985 by the US Air Force. Named the Mid-Infrared Advanced Chemical Laser or MIRACL, the unit was based on the ground and aimed at a surplus ICBM while in flight, which it disabled. It was later tested on drones simulating enemy cruise missiles. The tests were promising, but the beam has to be perfectly targeted and the burst sufficiently powerful at a given range to critically damage the target.

In a final test in 1997, MIRACL was aimed at an older USAF satellite that was soon to be deactivated. Operators waited until the satellite was in position, about 270 miles above the ground-based laser, then fired...but the laser did more damage to itself than the target, melting part of its own insides. After another test, MIRACL was set aside.

After the Soviet Union collapsed in 1991, efforts for military lasers were shifted to more portable, 'theater'-oriented programs. Elements of MIRACL were reconstituted in a program called the Tactical High Energy Laser, or THEL. A version of this, another deuterium fluoride laser, was mounted on a mobile vehicle and successfully shot down various test rockets of Russian design in 2001–2002.

Death rays made headlines again when the Boeing corporation and US Air Force released word of their YAL-1A airborne laser cannon. This managed to take all those nasty chemicals and control/navigation/targeting apparatus (and a full technical crew) and jam them into a recycled 747 airliner. The program used a chemical oxygen iodine laser, or COIL, described earlier in this chapter. The large chemical tanks were mounted amidships, with six laser modules in the back, a laser tube that ran most of the length of the aircraft, and an aiming turret on the nose with lenses to focus and direct the beam. This monstrous airborne laser-cannon was tested for years, with mixed results. It was ultimately deemed impractical due to high operating costs, the need to have a number of YAL-1's flying orbits around possible enemy launch sites, susceptible to anti-aircraft missiles and fighter jets, as well as the usual complications with lasers and weather conditions.

To conclude, many directed energy weapons (or DEW) have been tested. They are a lot tougher to make than even the best military minds had hoped, and there are many challenges to overcome. Bulk (huge mirrors,) energy requirements (enormous tanks of nasty chemicals,) guidance (all those electronics,) weather (a cloudy day can impede your shot,) and of course location (getting large masses into space, or flying them continuously in enemy airspace, is hard and expensive.) But there is one other type of weapon that deserves mention—the X-ray laser.

Lasers work in many frequencies. Visible light, infrared and ultraviolet are all contenders. But what if you wanted to make something that operated on a really extreme wavelength? How about X-rays? The SDI program at one point considered this idea, first suggested by Edward Teller, one of the fathers of the H-bomb. The problem was that an X-ray laser with a big enough punch to be an effective weapon needed a correspondingly big power supply. His solution was to use a nuclear bomb placed inside a large ball with a bunch of laser beam-creating rods sticking out of it. These rods could be oriented (to an extent) toward an entire fleet of incoming ICBMs. Inside the sphere would be a nuclear warhead. When it is ignited, energy from the brief flash created would enter the lasing rods, creating a *very* brief X-ray laser beam in numerous directions before the assembly destroyed itself. As Teller put it, 'A single X-ray laser module the size of an executive desk...could potentially shoot-down the entire Soviet land-based missile force....'

Despite Teller's optimism, in something like ten tests in the late 1970s and 1980s, the device never really delivered on its promise. The program was abandoned after the Cold War, but is still theoretically possible. Perhaps this was the ultimate expression of the death ray concept: it was hugely powerful, somewhat indiscriminate in its output and, like many great martial characters of myth and legend, self-destroying.

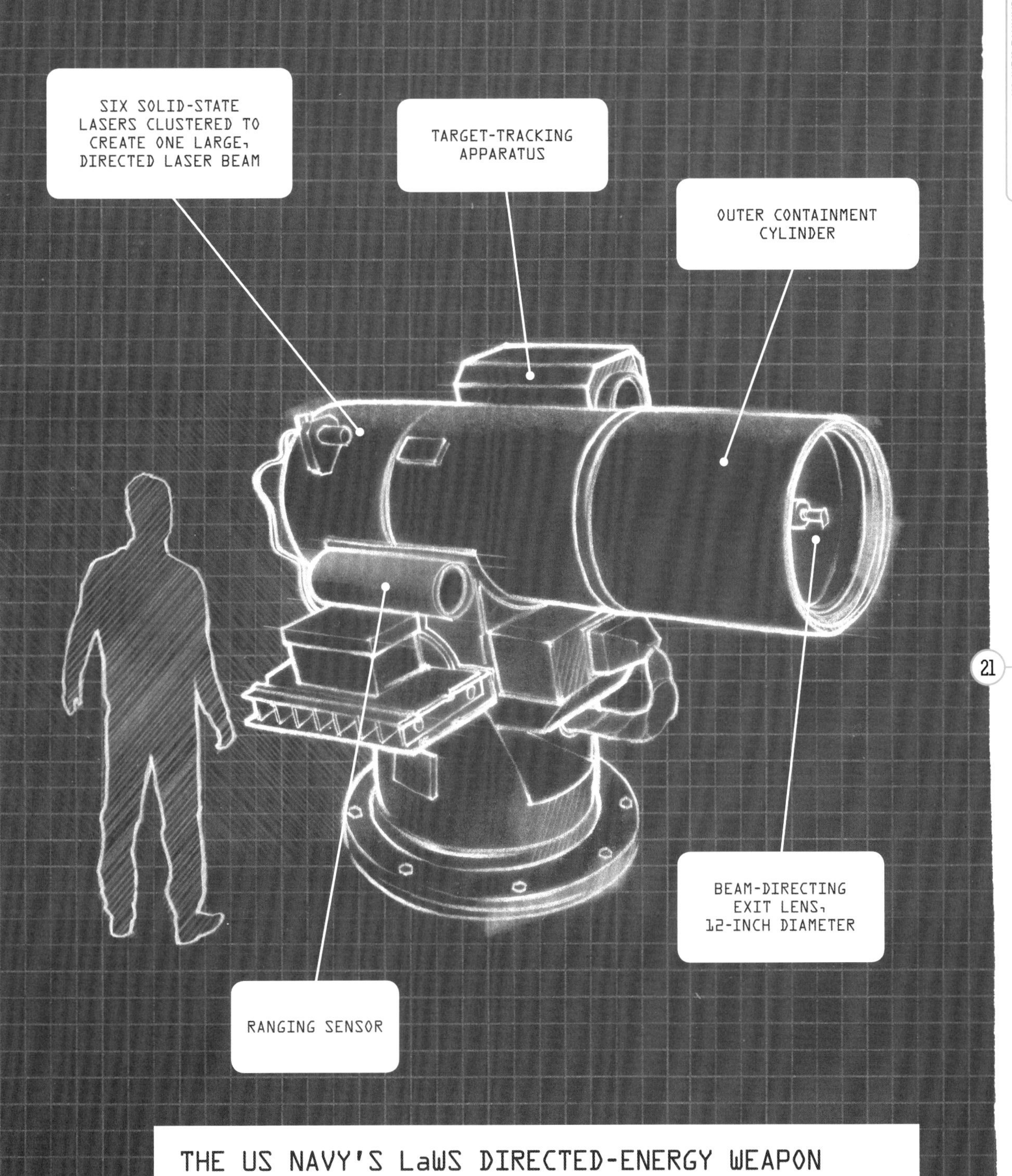

THE US NAVY'S LaWS DIRECTED-ENERGY WEAPON

LaWS is a ship-mounted ship-mounted solid-state infrared laser weapon. Power output is scalable for differing threats, to a maximum of 30 kilowatts. The weapon was deployed in 2014 on navy ship USS *Ponce*.

LASERS IN THE SKY: WELL, ALMOST. AFTER DECADES OF WORK, THE AIR FORCE DID SUCCEED IN DEVELOPING A HIGH-POWERED CHEMICAL LASER THAT COULD BE FLOWN TO ENEMY TERRITORY IN A CONVERTED 747 AND USED TO SHOOT-DOWN MISSILES AND, POSSIBLY, OTHER AIR ATTACK VEHICLES. BUT THE TECHNOLOGY WAS INCREDIBLY COMPLEX, AND THE LOGISTICS DAUNTING. THE PLANE HAD TO REFUEL BOTH ITSELF AND THE LASER FREQUENTLY, AND BAD WEATHER AND OTHER FACTORS COULD REDUCE ITS EFFECTIVENESS. THE AIRBORNE LASER PROGRAM WAS ULTIMATELY SHELVED.

SHIELDS UP!: CREATING A FORCE FIELD

Certain commands have become standard in science fiction: 'Shields up!'; 'Full power to the shields!'; 'Shields at 60 percent...'; 'The shields are down, sir!". Shields, force fields, deflector fields, energy screens, and 'repulsor' fields all refer to essentially the same thing: a protective energy field around a craft or building, which can deflect or absorb everything from energy blasts to heavy artillery, bullets, or meteor strikes. Starships, shuttlecraft, military bases, and exploration outposts have all used them in science fiction for decades. Early uses of the term as applied to spaceship defense predate *Star Trek* by decades. Shields were first powered-up back in the 1930s, E.E. 'Doc' Smith. Without shields, much subsequent plot-driven sci-fi entertainment would have been impossible.

PROTECTION FOR ALL

In many science fiction movies and novels, force fields are used to famous effect to protect starships, but have also been seen as personal energy shields, city-protecting electronic domes, fortress shields, shielding for open hatches on spacecraft, and dozens of other applications. How close are we to actually having this seemingly simple technology at hand? Not so far, in fact—but in a different form than you might expect.

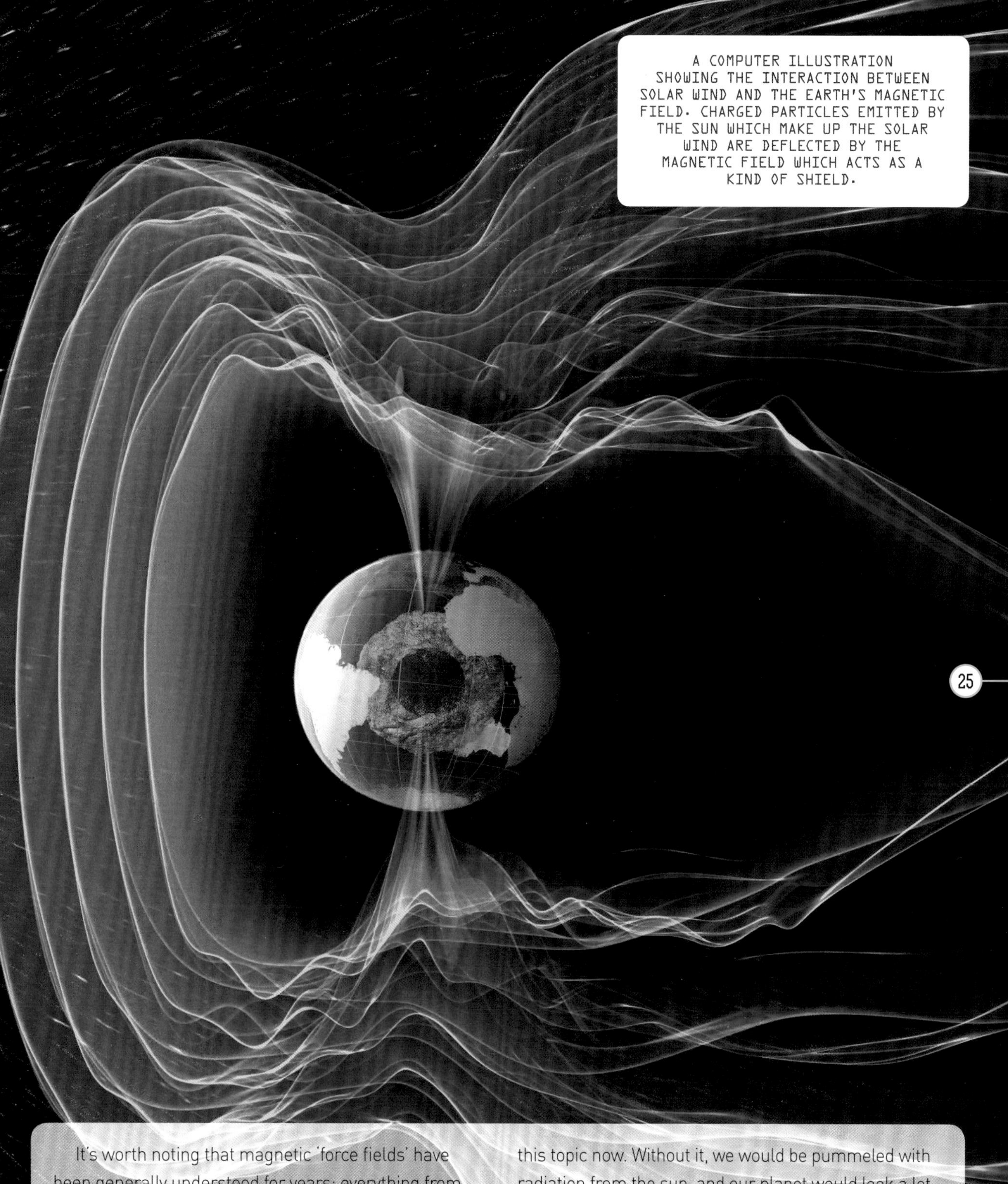

A COMPUTER ILLUSTRATION SHOWING THE INTERACTION BETWEEN SOLAR WIND AND THE EARTH'S MAGNETIC FIELD. CHARGED PARTICLES EMITTED BY THE SUN WHICH MAKE UP THE SOLAR WIND ARE DEFLECTED BY THE MAGNETIC FIELD WHICH ACTS AS A KIND OF SHIELD.

It's worth noting that magnetic 'force fields' have been generally understood for years; everything from the small lines of force created by a simple magnet to the gigantic magnetic field generated by the Earth is a type of force field. In fact, Earth's magnetic field is one of the primary reasons we even exist to be discussing this topic now. Without it, we would be pummeled with radiation from the sun, and our planet would look a lot more like Mars—which has only a weak magnetic field—and be essentially sterile on the surface. Earth's magnetic field traps some incoming energy (which would be harmful to life forms) in the atmosphere and channels the rest into the north and south poles of our planet. Handy! But magnetic fields, while incredibly useful in everything from shielding planets to generating electricity, have so far not been very effective in repelling conventional, explosive weapons. Unless you are shooting a bunch of magnets at a target, which is in turn pointing a large magnet at you, with the same poles (+ or -) pointing at each other, there won't be much effect (we have all seen how the like poles of magnets repel each other.) And of course, if the incoming magnets flip over, so that opposite poles are facing one another, the incoming magnets would simply accelerate the speed of the magnet, which would not be helpful. Of course, there is also the possibility that your bullets won't be made of ferrous materials and can blow right past my magnetic shielding, and so forth. Ultimately, other ways of manipulating magnetic forces may prove powerful in blunting attacks, but for now, not so much.

So how hard can it be to develop an energy shield against incoming threats? Pretty hard as it turns out. Unsurpisingly, our dear friend Nikola Tesla was sure he had it figured out in the early 20th century. He referred to the motive power behind his invention as 'scalar waves' or a 'scalar field.' These waves were said to contain magnitude only, not direction and, according to Tesla, if he projected scalar waves with multiple transmitters like radio waves, the area in which they crossed over each other would create an interference pattern. This would result in a force that would inflict vast destruction on objects entering that zone. While examined many times by many smart people, there is no evidence anyone has come close to getting scalar field weapons to work.

Tesla's laboratory

Tesla built this transmission tower and laboratory in Shoreham, New York in 1901. It was intended to be a wireless transmission station for sending voice and images across the Atlantic and to ships at sea, as well as a power transmission station. These were related to his ideas about wirelessly transmitting through the earth. Unfortunately, Tesla lost the support of J. P. Morgan, his primary backer, and it was abandoned in 1917.

More traditional approaches have been tried, but with only limited success. In a bit of inverse discovery, when nuclear weapons were first being tested it was discovered that they generated a huge burst of energy, known as an electromagnetic pulse or EMP. This had been predicted by Enrico Fermi, one of the principals in the creation of the atom bomb. Because of this, the military had shielded the electronics used to run and monitor the test. Fermi's theory was proved: this instantaneous release of energy in the form of magnetic and electric fields is death to any electronic equipment within a range that varies with the power of the weapon. While not a 'force field' in the traditional sense, the destructive power of an EMP burst is formidable—one of appropriate size would, for example, likely disable a large portion of the electrical grid of the country over which it was detonated. So while an EMP might not physically blunt incoming missiles or artillery, it could disable the targeting, firing, and navigational electronics in use. Of course, all modern weaponry intended for possible use in nuclear-warfare environments is 'hardened' against this kind of energy surge and would probably continue to operate.

While all the top military powers are surely working on some form of shielding tech at any given time (there are current rumors about China doing so,) the most public efforts are those of DARPA, the Defense

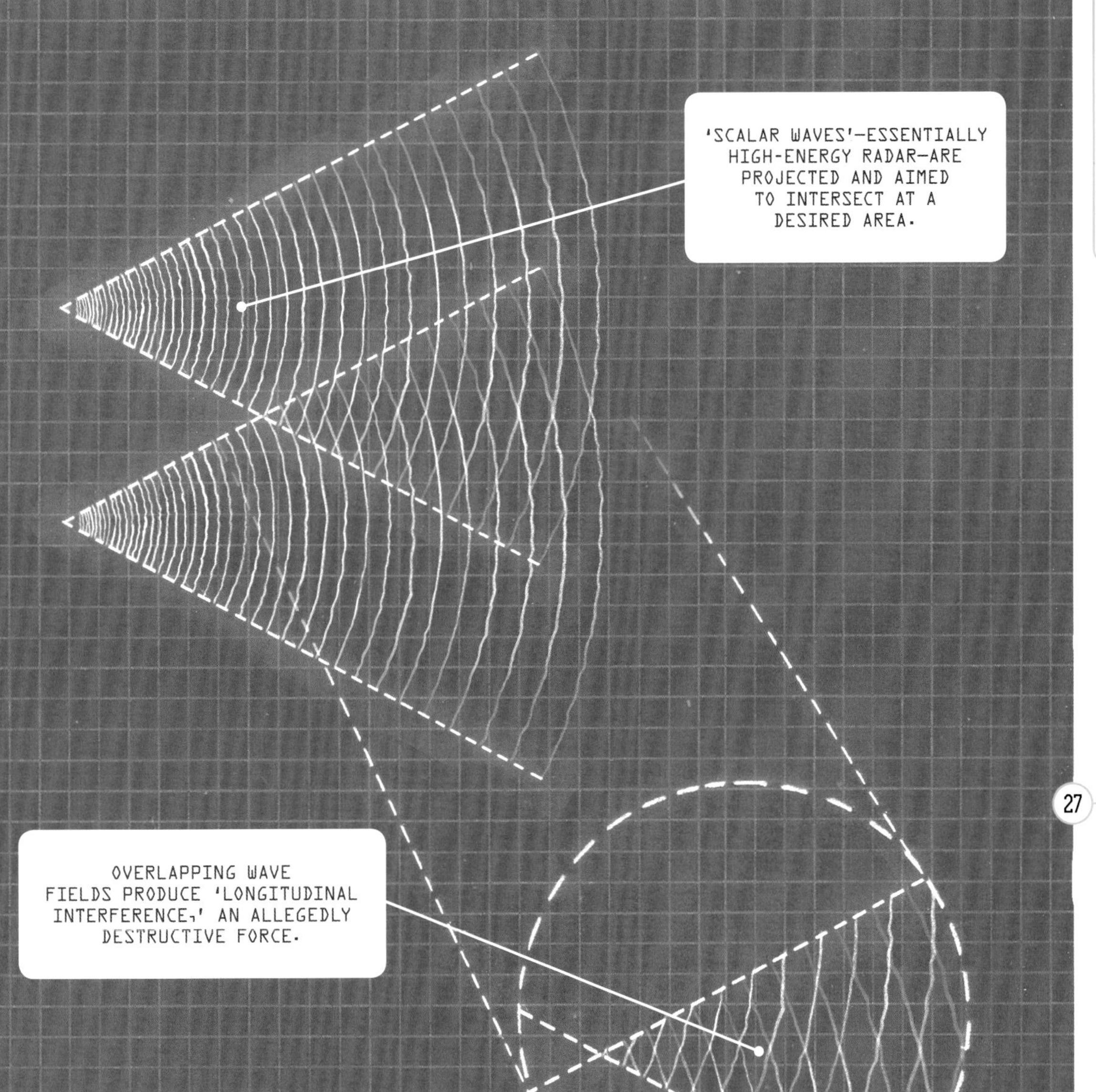

TESLA'S DETERRANT

Nikola Tesla came up with the idea of creating an interference with electromagnetic fields that he called 'Scalar Waves.' Two overlapping fields of properly created electromagnetic waves would, where they intersected, create a destructive force. As far as is known, it was not tested.

Advanced Research Projects Agency, in the United States. You will see their name pop up frequently in this book, because they are the funders of so much that is crucial to the US tech research effort in warfare and civil defense. DARPA's mission is to push the limits, explore new and advanced technologies, and assure that there are as few surprises in store for the US military as possible.

One of their more recent accomplishments was invented by aerospace and defense contractor Boeing, who recently patented a device that uses ionized air to protect 'soft targets'—in other words, living soldiers—from the shock created by explosions. Entitled 'Method and system for shockwave attenuation via electromagnetic arc,' the system is designed to minimize injury to soldiers within a vehicle that has been exposed to an explosion. That explosion could range from an artillery shell to a roadside Improvised Explosive Device (such IEDs were particularly dangerous in Iraq and Afghanistan.) Normal armor can protect the occupants from the heat and destructive force of the explosion, as well as shrapnel, but not the shockwave.

The system is fairly simple in theory—but, as is so often the case, more difficult in execution. The patent drawing shows an example of the system mounted on an armored US Army Humvee. A sensor detects an explosion of some kind, be it shell, grenade,, or IED. It then creates a pocket, or 'shield,' of extremely hot, ionized, and dense air—a plasma—to that side of the Humvee, using a laser or electrical arc. The system would not actually disable a projectile. Rather, it is designed to minimize the effects of the resulting shockwave via 'reflection, refraction, dispersion, absorption and momentum transfer...' Work is clearly still ongoing, but if it can be made to work, the system would mean not just fewer deaths, but far fewer injuries to the body and brain.

The British Ministry of Defence has its own version of DARPA, called the Defence Science and Technology Laboratory. These folks have developed about the closest thing to actual sci-fi shields yet. It involves a new type of armor plating that can be applied to tanks, armored personnel carriers, and other kinds of vehicles, and which can actually *repel* medium-sized munitions.

The new armor uses a powerful electrical charge to create a strong electromagnetic field across the surface

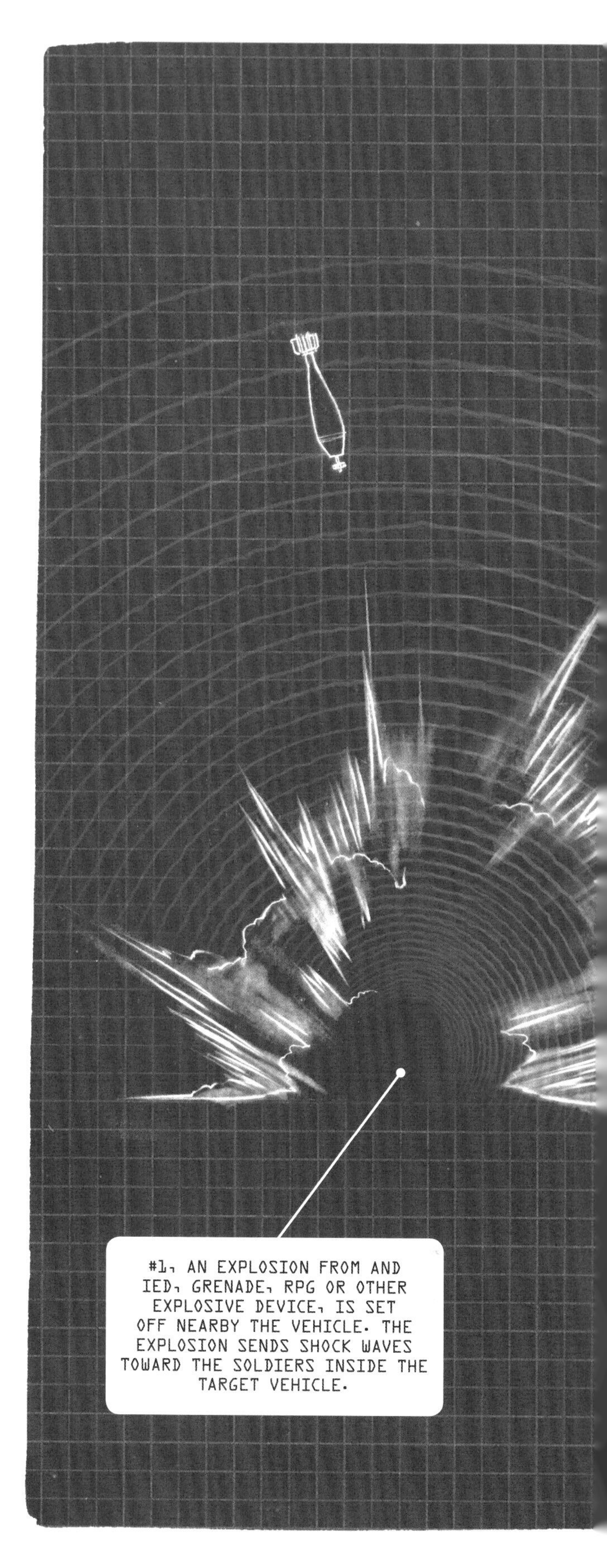

#3, A PLASMA BALL IS CREATED BY INSTANTANEOUS IONIZATION OF A POCKET OF AIR, DEFLECTING THE SHOCKWAVES.

#2, ELECTRICITY IS STORED IN VEHICLE-MOUNTED CAPACITORS. WHEN SCANNERS SENSE AN EXPLOSION, THE CAPACITORS DISCHARGE INSTANTLY, POWERING A LASER OR MICROWAVE GENERATOR AIMED TO THE SIDE.

INSTANTANEOUS DEFLECTOR

A system recently invented by Boeing uses a high-energy ball of plasma to deflect the energy of a nearby explosion.

of the metallic armor plate. It does this via 'supercapacitors,' an ultra-powerful version of the same kind of capacitor that powers the strobe in your smartphone's camera. Like the Boeing device already discussed, this is a 'momentary' shield, which would require tracking of an incoming threat and an instantaneous activation to be effective. Capacitors discharge extremely quickly and the shielding would last for only a fraction of a second. They would then need to recharge before being used again.

The advantages of such a system are obvious: reduced thickness and weight of vehicle armor. And in time, such tech might even be applicable to personal body armor in the form of an exoskeleton-type full-body armor, providing that the person inside is protected from the strong shock. Downsides include the inability to repel non-metallic threats, the need to discern large threats from small ones, and recharge time—rapid-fire threats could overcome the system unless it can be recharged instantaneously. But it's a start.

One of the biggest destroyers of military vehicles are rocket-propelled grenades (RPGs). These relatively small and highly portable weapons create a jet of molten metal upon impact, which can penetrate steel armor up to a foot thick. The new electric armor can stop such munitions from ever pentrating the protected vehicle. It

would also be effective against traditional armor-piercing rounds and other munitions. And the traditional armor plating protects against small-arms fire, so the electric armor would need to be 'tuned' only to activate for larger incoming threats.

Other versions of this idea have been floated, and sometimes experimented with. One idea is to create a 'counterblast,' a shaped explosion that fires in the direction of the incoming force, therefore reducing its potential for damage. Another is to employ large amounts of quickly blasted water to attenuate the incoming forces. Still others heat quickly deployed clouds of specific gases—argon or helium, for example—into a plasma, resulting in an effect similar to Boeing's device. None seems quite as practical as Boeing's (which instantly heats a chunk of the atmosphere,) and something with the potential for multiple-uses in quick order would probably be desirable. But the amount of interest in these ideas is proof positive that shockwave 'shielding' is certainly a desirable goal.

This is the closest thing to deflector shields that world militaries are admitting to, but who knows what other technologies might be emerging behind sealed (and possibly force-fielded) doors?

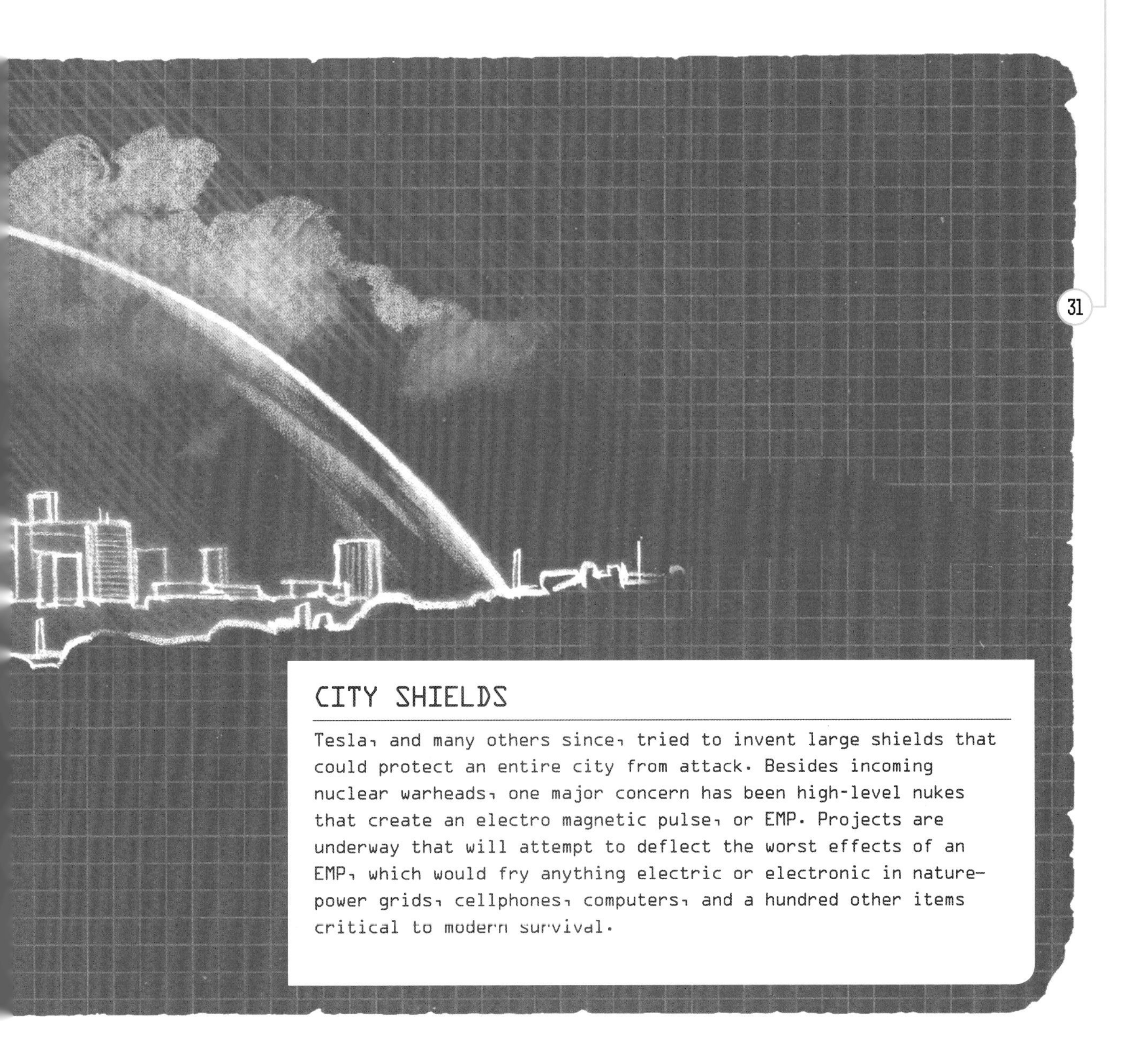

CITY SHIELDS

Tesla, and many others since, tried to invent large shields that could protect an entire city from attack. Besides incoming nuclear warheads, one major concern has been high-level nukes that create an electro magnetic pulse, or EMP. Projects are underway that will attempt to deflect the worst effects of an EMP, which would fry anything electric or electronic in nature—power grids, cellphones, computers, and a hundred other items critical to modern survival.

SWORD OF HEAT: USING LIGHT SWORDS

A light sword is the ultimate accessory for a science fiction warrior, and an essential part of the universe of *Star Wars*. But light blades had far earlier origins.

INTERESTING ORIGINS

'Energy swords' were seen in science fiction in the 1930s, with the first appearance possibly in Edmond Hamilton's *Kaldar: World of Antares* (1933.) By the time a light sword showed up in the first *Star Wars* movie in 1977, it had been refined into an electronic beam of roughly four feet, emanating from a metal handle. In reality, they were Heiland flash guns, akin to oversized flashlights. They carried D-cell batteries inside and normally had a six-inch reflector at the top, pointing forward if the device was held vertically. Inside the reflector would be a flashbulb—a one-time use lightbulb filled with magnesium wire that, when hit by an electrical impulse from the batteries, would emit a brief pulse of blinding light as the wires burned up. Photographers swore by them for decades, as they were the only portable way to instantly light dark settings portable strobe lights had not been invented yet.

In the early *Star Wars* films, the energy bolt that emanates from the light sword's handle was rotoscoped, hand-painted frame by frame over a reflective pole. The later films used computer graphics to stunning effect, with appropriate sound effects. A variety of light sword stand-ins can be bought, from cheap plastic toys to high-priced reproductions made from metal with exotic and often complex mechanisms to simulate the beam. A real light sword, though, is a totally different matter.

This is where the engineering gets a bit sticky. Light swords are not just lasers or plasma beams... they are bolts of energy able to deflect blows from other light swords, reflect blaster bolts, and cut through solid objects like butter. So they need to have some kind of mass, polarity and, presumably, high temperature. These requirements turn out to be a tall order, even with a roomful of energy generation equipment and state-of-the-art lasers.

The most obvious property of the light sword is simply the energy beam, a brightly-lit spike of energy about four feet long. This probably sounds as if we are talking about a laser, except that for one huge and vexing problem...the beam *stops* at a defined length. Lasers and other beams of high-powered light extend basically into infinity (or until something, like a cloud or solid object, stops them.) In fact, the whole idea behind a laser is that of 'coherent light.' The light beams are generated and treated in such a way that the frequencies align, staying sharply focused and transporting whatever power they contain over great distances if not impeded by solid objects, dust or clouds. So there is no known way to stop a hand-held laser beam at four feet.

What about other forms of power? Bolts of electricity are a nice candidate, as they have a length determined by the amount of power generated. But a four-foot bolt of lightning would need a whole lot of power to create—like a backpack the size of a refrigerator—and would need big, expensive capacitors to store up the electrical potential which would discharge in an instant and need a lengthy recharge cycle.

Electricity has other annoying traits. While the electrical spark, or bolt, can be generally controlled regarding length, it tends to move in whatever direction it wants, so long as it is toward something that will conduct electricity. This would include the person holding the light sword, rendering the weapon just as dangerous to the person wielding it as to the enemy.

Plasma is another option for creating a light sword. A 'plasma torch' jet can be formed by expelling electricity-charged gas—oxygen, for example—out of a tube. The result is a brightly-colored flame of extremely high temperature. Plasma torches are commonly used in industrial settings to cut through metal, so the resulting flame is really hot. And it has a defined length (though more like six to nine inches than four feet.) And, perhaps as important as all the rest, a very cool, bright glow.

But alas, the cutting process is slow (at least with hard objects like metal,) plasma beams do not bounce off each other, they do not reflect blaster bolts (at least not lasers,) and once again the power requirements are damning—you would need a cable from your light sword to a high-amperage wall plug, which would make dueling a little impractical. Currently, they would also require a constant flow of coolant to avoid melting, which is also not conducive to striking a good fighting stance

There is one small light on the horizon. In 2013, researchers at the Massachusetts Institute of Technology announced that they had 'accidentally' discovered a form of energy similar to *Star Wars*-style weapons. The Harvard-MIT Center for Ultracold Atoms discovered a form of light that acts as if it has physical mass. When the photons (light particles) generated by their device come up against each other, they bounce away, or deflect one another. They don't have the high-temperature cutting effects of a hot laser, nor do they make the cool electronic noises a good light sword does. But it's a start.

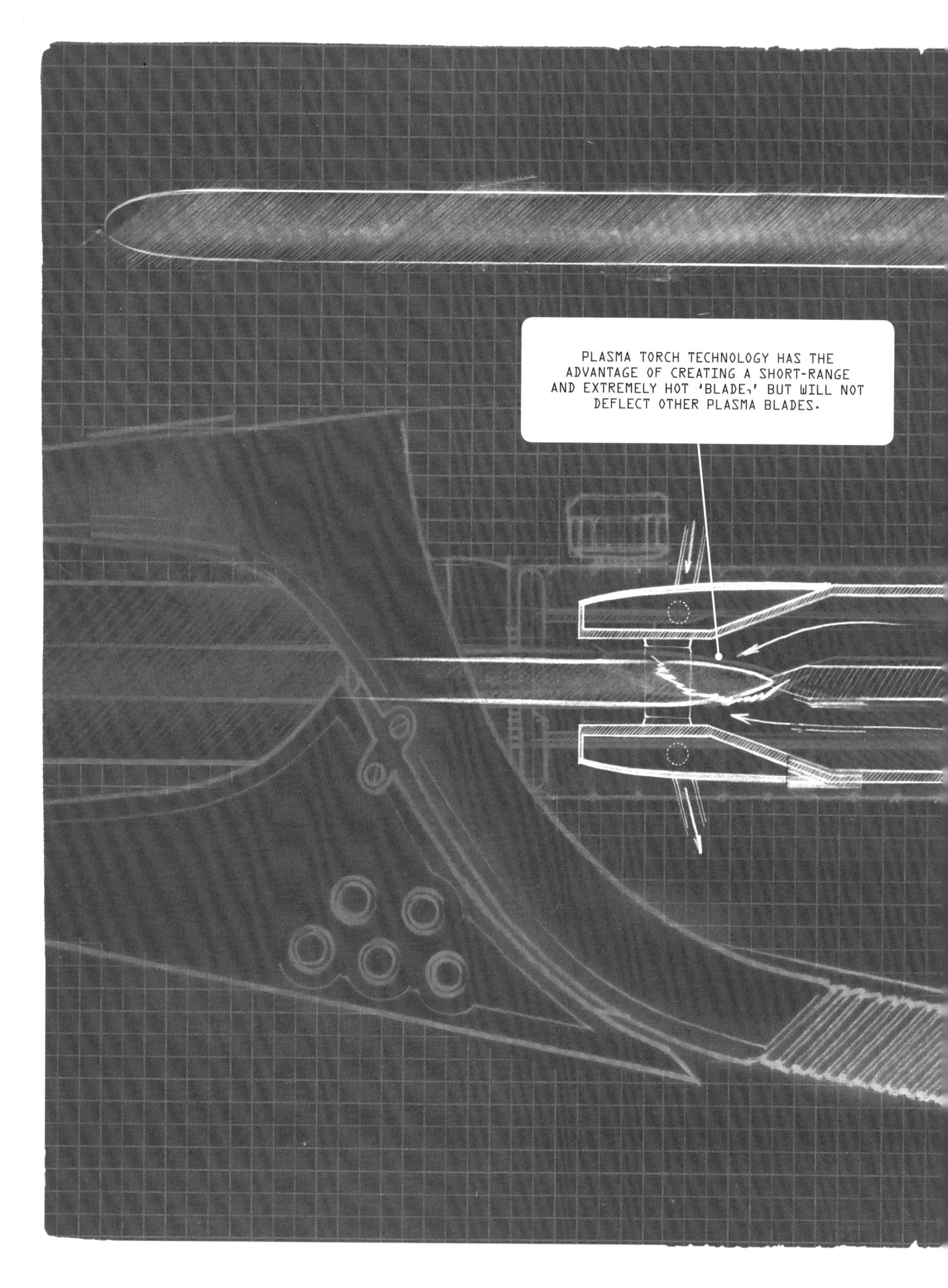
PLASMA TORCH TECHNOLOGY HAS THE ADVANTAGE OF CREATING A SHORT-RANGE AND EXTREMELY HOT 'BLADE,' BUT WILL NOT DEFLECT OTHER PLASMA BLADES.

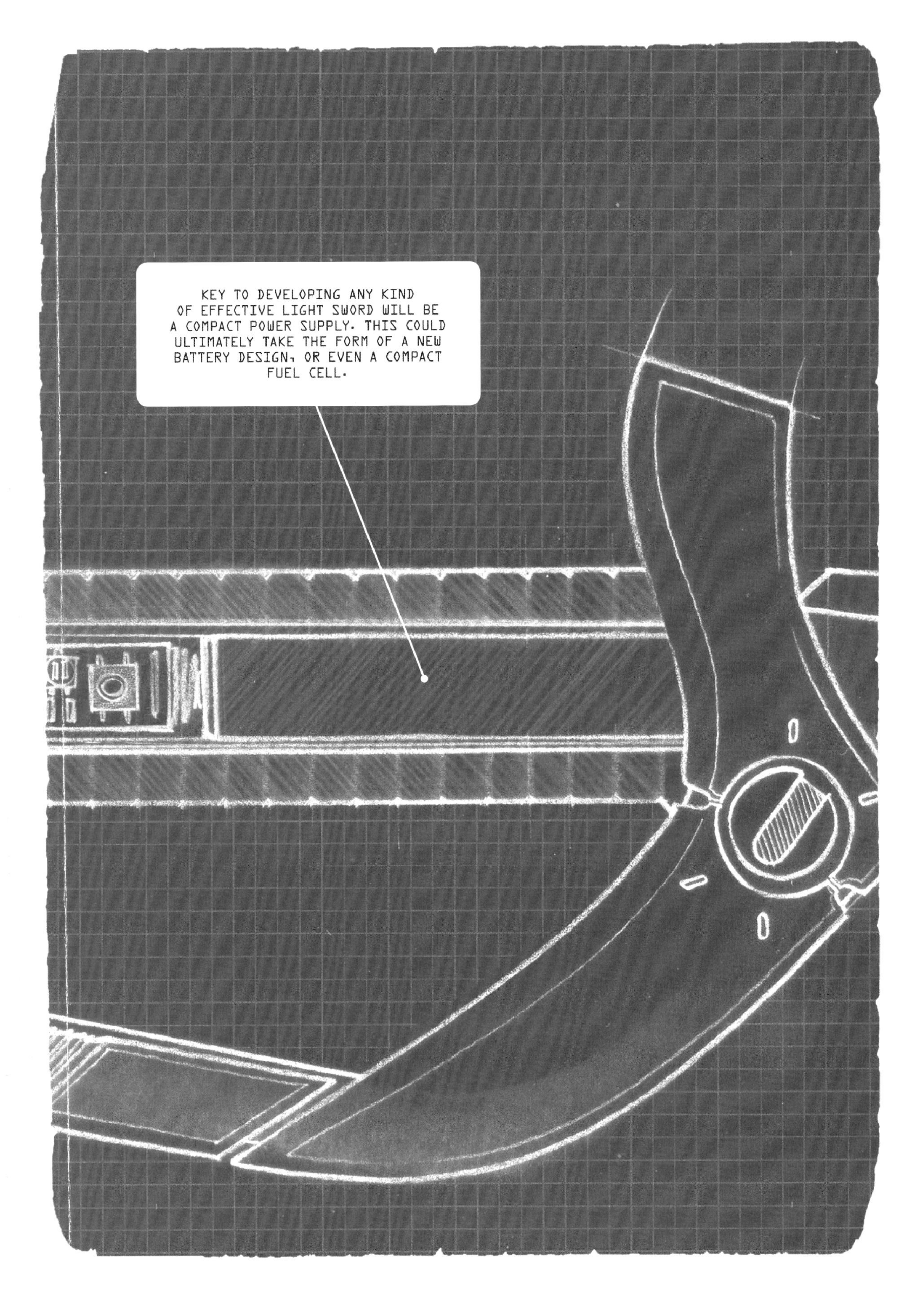
KEY TO DEVELOPING ANY KIND OF EFFECTIVE LIGHT SWORD WILL BE A COMPACT POWER SUPPLY. THIS COULD ULTIMATELY TAKE THE FORM OF A NEW BATTERY DESIGN, OR EVEN A COMPACT FUEL CELL.

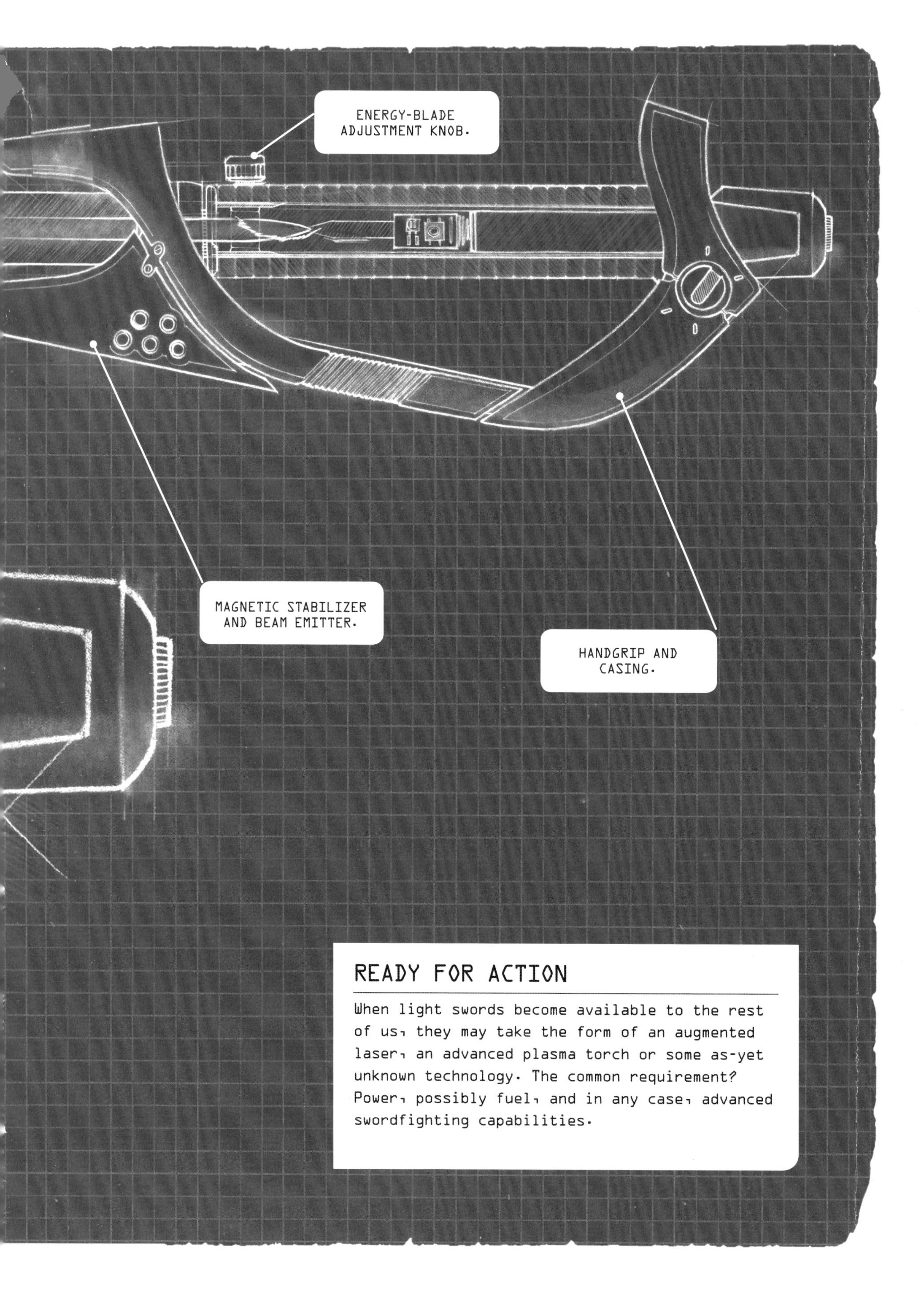

READY FOR ACTION

When light swords become available to the rest of us, they may take the form of an augmented laser, an advanced plasma torch or some as-yet unknown technology. The common requirement? Power, possibly fuel, and in any case, advanced swordfighting capabilities.

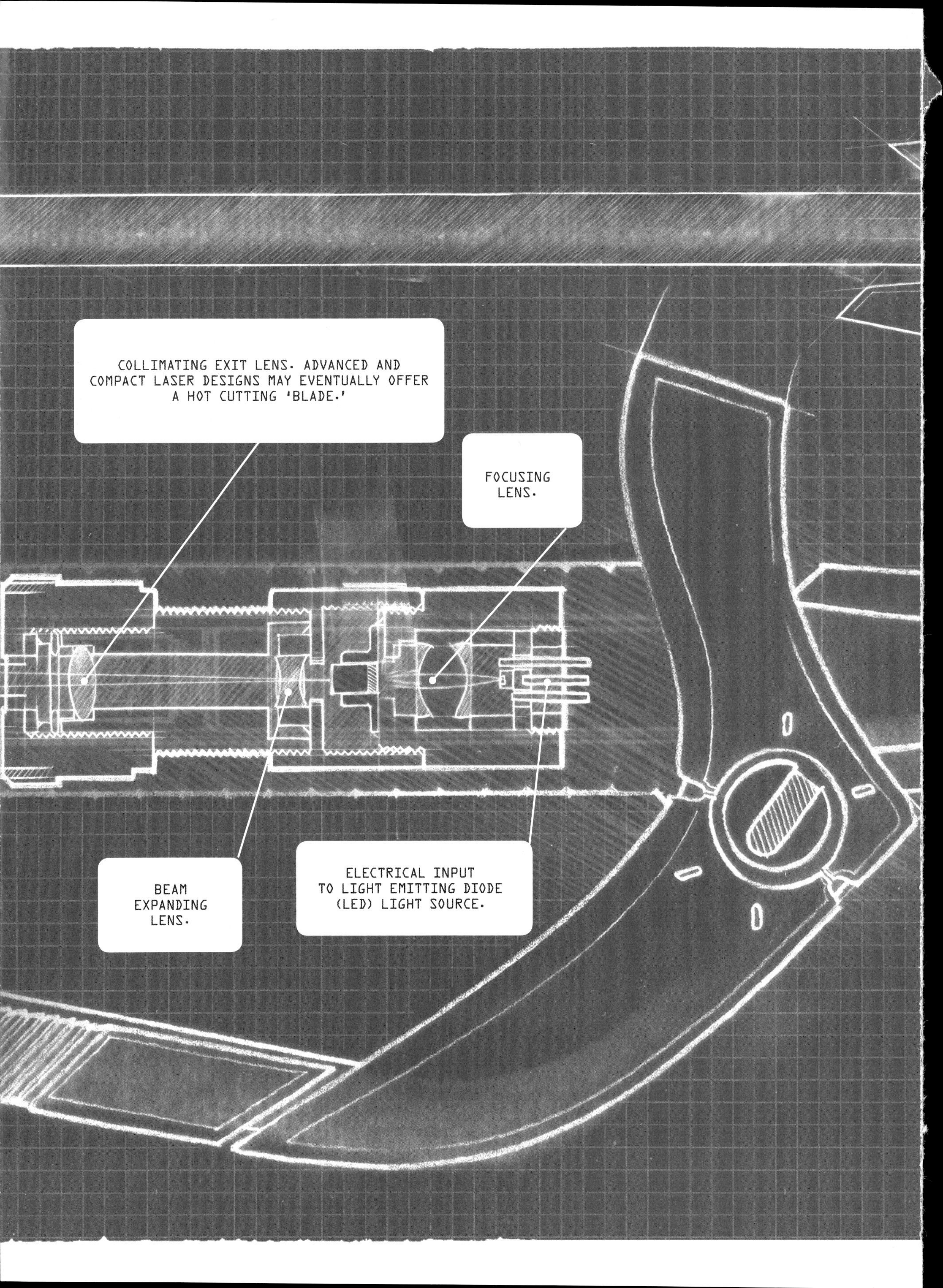
COLLIMATING EXIT LENS. ADVANCED AND COMPACT LASER DESIGNS MAY EVENTUALLY OFFER A HOT CUTTING 'BLADE.'
FOCUSING LENS.
BEAM EXPANDING LENS.
ELECTRICAL INPUT TO LIGHT EMITTING DIODE (LED) LIGHT SOURCE.

JUDGMENT DAY: THE RISE OF AI

There is a simple but chilling exchange in the 1968 movie *2001: A Space Odyssey* between astronaut Dave Bowman and the HAL 9000 computer. HAL is attempting to kill the last living astronaut aboard the Jupiter-bound *Discovery* to preserve the integrity of 'the mission.' When *2001* hit the screens, computers were still room-filling rarities; nonetheless, AI quickly became the perfect fictional adversary, cementing itself into our minds by HAL as the ultimate villain, and onlysometimes ally. Many other movies, TV shows, and books followed with AI in lead roles, from the smooth-talking KITT ('Knight Industries Two Thousand,' an AI computer-in-a-car) in the TV series *Knight Rider* to the bent-on-genocide Skynet in the *Terminator* films.

SUPER INTELLIGENCE

AI's cold efficiency is the very thing that scares people like Elon Musk and Stephen Hawking. Musk recently said that he saw AI as the 'greatest existential threat' to the future, and that it was time for some 'regulatory oversight at the national and international level.' Bill Gates, who had a hand in the creation of basic AI, agrees with Musk. 'I'm in the camp that is concerned about super intelligence,'he has said, referring to unrestricted AI development.

And then there is Stephen Hawking, who believes that success in AI would be the most significant event in human history. However, he also recognizes that it might be one of the last, unless we learn more about the risks involved and how to avoid them. Professor Hawking goes on to point out that the militaries of the world are already placing some decision-making power in the hands of AI.

So, if AI is that risky, perhaps we should look more closely at what it actually is. Dr. John McCarthy, who was a professor of computer science at Stanford, defines it as 'the science and engineering of making intelligent machines, especially intelligent computer programs. It is related to the similar task of using computers to understand human intelligence, but AI does not have to confine itself to methods that are biologically observable.' Even intelligence itself, though, is a slippery concept. McCarthy explains that:

Intelligence is the computational part of the

ability to achieve goals in the world. Varying kinds and degrees of intelligence occur in people, many animals and some machines...[t]he problem is that we cannot yet characterize in general what kinds of computational procedures we want to call intelligent. We understand some of the mechanisms of intelligence and not others.

The ultimate test of AI has long been known as the Turing test, named after famed mathematician and futurist Alan Turing. The key test as to whether a robot or other AI machine can pass as a human (see Chapter 14,) McCarthy argues that if a machine could successfully pretend to be a human to a knowledgeable observer then it should be considered as intelligent. 'This test would satisfy most people,' he remarks, 'but not all philosophers.' He concluded by warning, 'It turns out that some people are easily led into believing that a rather dumb program is intelligent.'

This was written in 2007, but still applies in general terms. So is something like Apple's Siri true AI? It acts like a person and feels like a person, but in this case we clearly know that it is not. But in a Turing Test setup, without hearing the voice or knowing the source, we would likely think that we were corresponding with a person. So by that definition, Siri is true AI.

But while Siri (and its ilk) can at times be convincingly human, the programs can be fooled and often make errors that a ten-year-old would probably spot instantly. So it is wise to bear in mind that criticial decision-making authority or running the national defense should be reserved for humans. Acting human does not equal *being* human.

On the other hand, there are times when software makes better decisions than humans do. When IBM's Deep Blue computer beat the then-world champion chess player Garry Kasparov, it was making superior decisions on how to play chess—pure, cold logic.

Modern AI is less monolithic than originally conceived; each application is highly specialized, and we probably won't be coming up with a universal cyberbrain anytime soon. But in some applications of AI software, the logic applied to the task becomes fraught with ethical issues that are difficult to program. If one of Google's self-driving cars is faced with a split-second decision in which lives are at stake—for example, in a potential collision scenario, does it: (a) collide with the car ahead and risk 'x' level of injury to the occupants of both vehicles; or (b) veer to the curb and risk the death of a pedestrian there? This is an oversimplification, but you can see the problem. Do we want software to make those kinds of decisions? On the other hand, whatever choice is made will be far faster and often more reliable than the human equivalent. A computer's judgment is almost certainly better than that of a drunk driver.

Like it or not, you are surrounded by AI entities already. Siri (and its Android equivalents) already drive much of our decision-making. Amazon's new Alexa cloud-connected voice-response units are listening to their users' households 24/7 and not only suggesting what music selection they might want to make next, but also preparing potential shopping lists based on their 'preferences.' Of course, that's been going on in the background of your web use for years; previous searches and purchases have been dutifully recorded and even emails mined to determine what kind of advertising might be 'helpful' to you.

AI seems to be primarily driven by two forces: consumer preference on the retail end and market/ military needs. It makes sense, both money and survival are good motivators, and both have succeeded at spurring AI development handsomely.

On the retail/consumer level, much work goes into social intelligence, which can be a kind of Turing Test. How can we make this software seem more human? This includes understanding and comprehension on the input side, where the machine needs not only to comprehend the direct input—what is stated by the user—but also to try to intuit emotional and other unspoken needs if possible. Then, both subtle and direct simulations of emotion and intelligence are applied on the output side. The results are mixed; Siri is great for telling you where the nearest pizzeria is, but not so good at working out why you're going there (a romantic dinner, a birthday party, or just to comfort eat after a breakup?). It takes more intelligent software than we currently have to do that.

Even so, AI can use methods other than real understanding to lure us into a state of trust. The Japanese researchers discussed in Chapter 14, the fembot-makers, have put a lot of work into making their pretty young robots act as if they like and understand us—lips curve into innocent smiles and heads tilt alluringly—but the machine does not *understand* you any better than Siri. However, given the multiple visual signals you are receiving that this is a 'person' looking at you and reacting in an emotionally complellin fashion, you are more inclined to give trust.

Networking computational requirements may help refine AI comprehension. The bigger the database, and the more instantaneous comparisons that can be made, the better the understanding will be. When your nurse-bot at the hospital has access to the cloud and its vast reserve of recorded human responses to given input, it is more likely to give a convincing and accurate response. Increasing data-points of human behavior and responses will help to offset some of the limits of the local processor.

So the question then becomes: is this *real* intelligence, or simply a better simulation? And that is the huge question for the future of AI. Does it really understand its decision-making, or is it simply responding with a convincing answer, padded with appropriate emotional cues? That said, for most humans, personal experience combines with learned knowledge to guide decision-making, and how different is that from what the machine is doing?

The key to the future of AI is software that *learns*. Google's search engine is transforming into a learning-based system as you read this. Online searches have long counted on algorithms written by humans, with human-created rules, to exhaustively research your query. These rules have dominated the results you got (along with some expensive placement preferences from high-paying advertisers.) But in 2015, Google moved into previously untrodden territory by implementing 'deep learning' with neural networks. These networks use vast amounts of computing to simulate the basic reasoning of the human brain, and the result has been that the trusted search engine now gives much more relevant answers to queries.

The scary thing is that even the programmers, when pressed, will admit that they do not fully understand how this system works. Neural networks learn as they operate, and apply that knowledge to what they do next. The best way to understand how these human-generated (so far) algorithms are learning is to examine them with other computer-based tools. It becomes a recursive process, and at some point, improvements to the software will be made by other software, and that is where some people begin to get nervous.

Computers taking decision-making away from humans has been a science fiction standard for decades. It was a key element in movies like *The Terminator*, where the computer defense system called Skynet became self-aware and decided that humans must be exterminated. So far, we have control over how this process works...but that is by no means how the future will play out.

Defense is a key area for the implementation of AI. DARPA's interest in the field is intense and multifaceted. Cyberattacks from places like China, Russia, and to a lesser extent the Middle East are a growing concern, and the stakes are enormous. We often think of defensive AI in a battlefield context only. And though that's important (because the side that wins the cyber-battle will likely win the war), there is more at stake here: national energy grids, tracking and response systems, space-based assets like satellites and communication, most banking, cargo delivery and tracking, medical records, and much more are all dependent on computer control

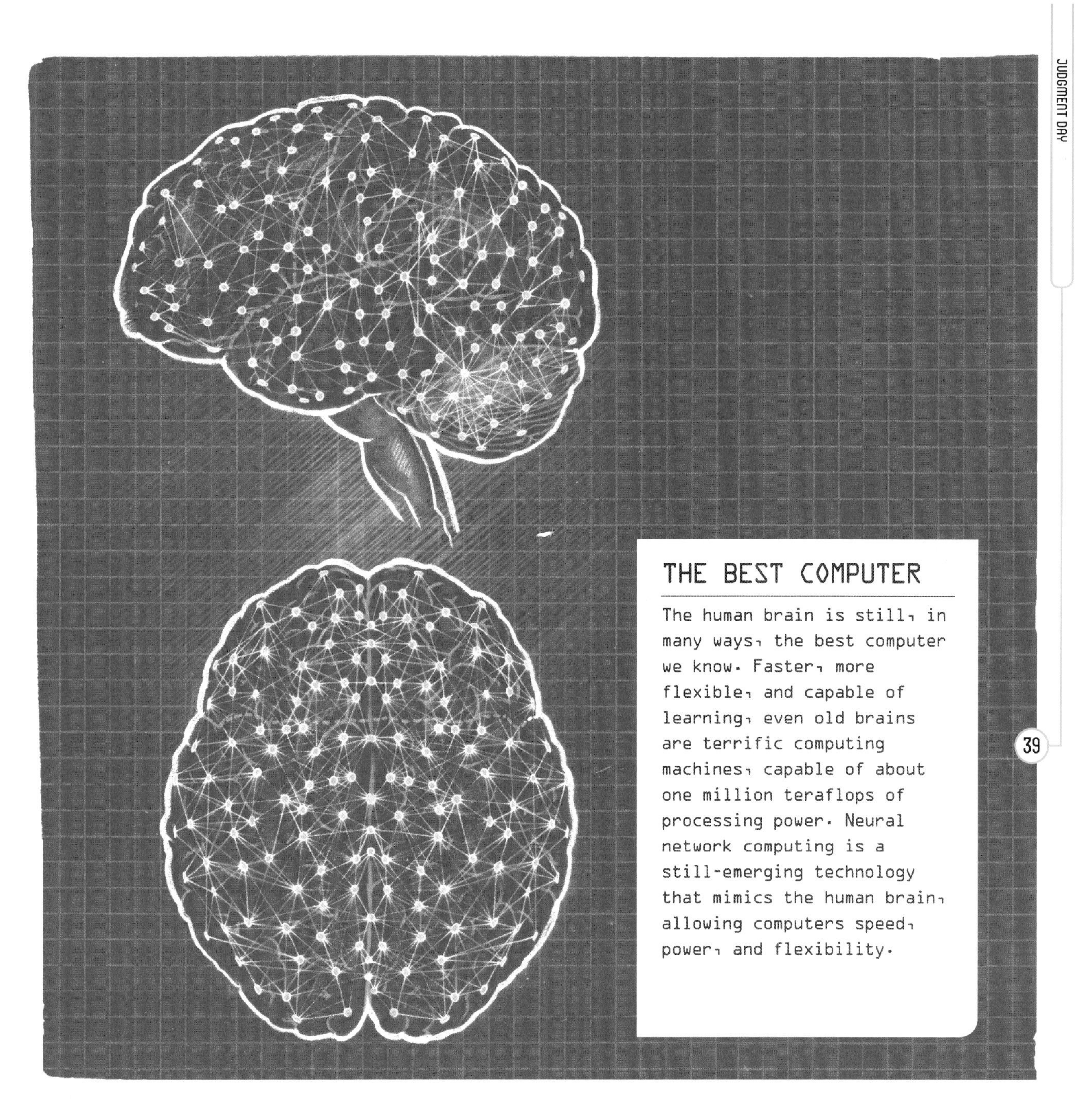

THE BEST COMPUTER

The human brain is still, in many ways, the best computer we know. Faster, more flexible, and capable of learning, even old brains are terrific computing machines, capable of about one million teraflops of processing power. Neural network computing is a still-emerging technology that mimics the human brain, allowing computers speed, power, and flexibility.

and software, and all are susceptible in some way to hacking.

Even so, the most immediately critical concern to the military is the global battlefield (including computerized control of local operations.) The Pentagon wants to standardize as much of the many different kinds of software powering their weapon systems as possible, but the more you network, the more you invite hacking by the enemy. Wireless control makes systems much more vulnerable. This is an area to which DARPA wants to apply the power of advanced AI.

AI systems currently used to safeguard wirelessly controlled battlefields scour the electromagnetic spectrum for enemy signals and analyze them by applying machine learning to predict what the enemy might do next. They then use various countermeasures to jam or defeat the intervention, protecting the command and control networks used to direct weapons systems.

But what about more direct applications of AI? Should robots be allowed to kill? So far the United States claims that all decisions regarding the use of lethal force are left in human hands, and that the smart computers merely present options for human commanders to choose from. What might be in the works within foreign militaries is less clear; many have not signed the limited agreements in place. And even with the military powers that leave human decision in the chain of AI command, the pace of battle can, at some point, overwhelm those human operators. The more advanced the tech of the enemy, the faster you need to respond. This may make the move to fully autonomous, automated, AI-run systems inevitable.

So where do we stand on truly 'smart,' free-standing, inference-reading AI? DARPA has been funding a program called SyNAPSE in an attempt to create a supercomputer that can think at the level of a mammalian brain, but this has not yet come close to human intelligence. Human brains are still many hundreds of thousands times more powerful and complex than the most advanced computers. The best SyNAPSE can hope for is a dogs-and-cats level of intelligence—but even that is challenging. This kind of neural networking has come up against the limits of traditional microprocessors, and a new approach called neuromorphic computing that more closely mimics how human brains work is being pursued.

This new computer architecture is a move away from the conventional model—serial (or linear) processing, which is how today's computers work, passing information back-and-forth between a microprocessor and electronic memory to accomplish calculations. An update of this is cloud computing, but

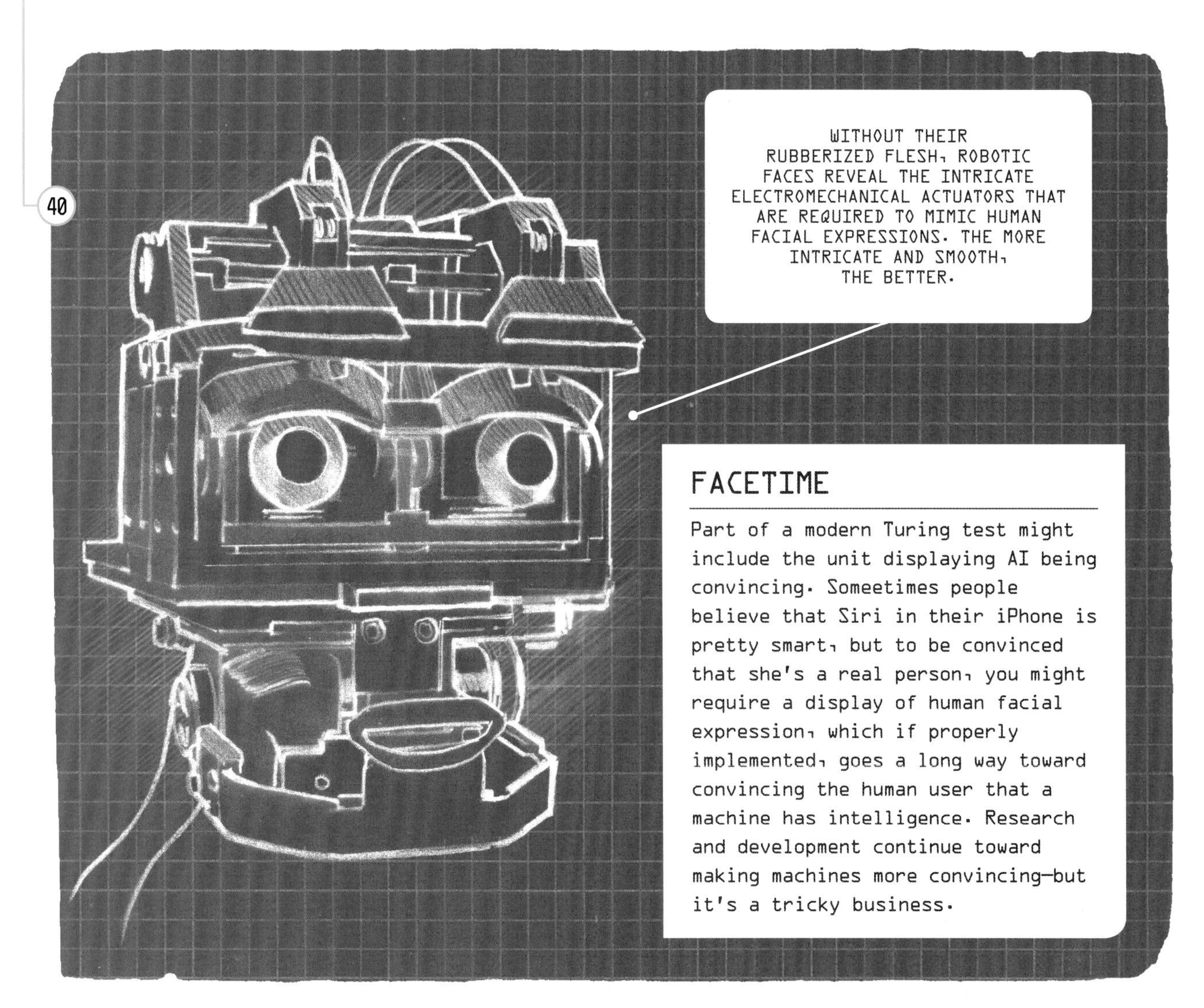

WITHOUT THEIR RUBBERIZED FLESH, ROBOTIC FACES REVEAL THE INTRICATE ELECTROMECHANICAL ACTUATORS THAT ARE REQUIRED TO MIMIC HUMAN FACIAL EXPRESSIONS. THE MORE INTRICATE AND SMOOTH, THE BETTER.

FACETIME

Part of a modern Turing test might include the unit displaying AI being convincing. Someetimes people believe that Siri in their iPhone is pretty smart, but to be convinced that she's a real person, you might require a display of human facial expression, which if properly implemented, goes a long way toward convincing the human user that a machine has intelligence. Research and development continue toward making machines more convincing—but it's a tricky business.

this simply calls out to more powerful remote systems that are working within the same framework. Neuromorphic processors will operate in parallel. This will enable them to learn much more quickly, using less power and wasting less energy as heat, the bane of traditional processors.

At the moment, these systems are still limited in what they can accomplish. But rapid learning of simple processes has been demonstrated. Test robots can begin a task (like cleaning the floor,) use cameras to observe how a human does it, and then alter their actions to incorporate what they have learned. It is not difficult to imagine an ATLAS-style battlebot (the DARPA two-legged search-and-rescue robot we will meet in Chapter 14) embedded in a human platoon as part of military training, then using the knowledge gained to engage a real enemy.

In the end, AI will be what we make of it. Not all nations (or their militaries) will agree to play by the same rules. A few years ago, the UN held breakout conferences at their Convention on Certain Conventional Weapons; it was essentially a referendum on killer robots. The general feeling was in opposition to the increased use of robotics in combat. It is simply asking too much of autonomous computing and software to make life-or-death decisions. But given what we have seen even in civilian R&D, AI will increasingly do our fighting, policing, and surveillance.

In the final analysis, the best we can hope for is strong oversight and regulation of AI and the technologies it inhabits and augments. Until someone can program *empathy*, or at least sound military judgement, the end-point alternative is just too chilling to contemplate.

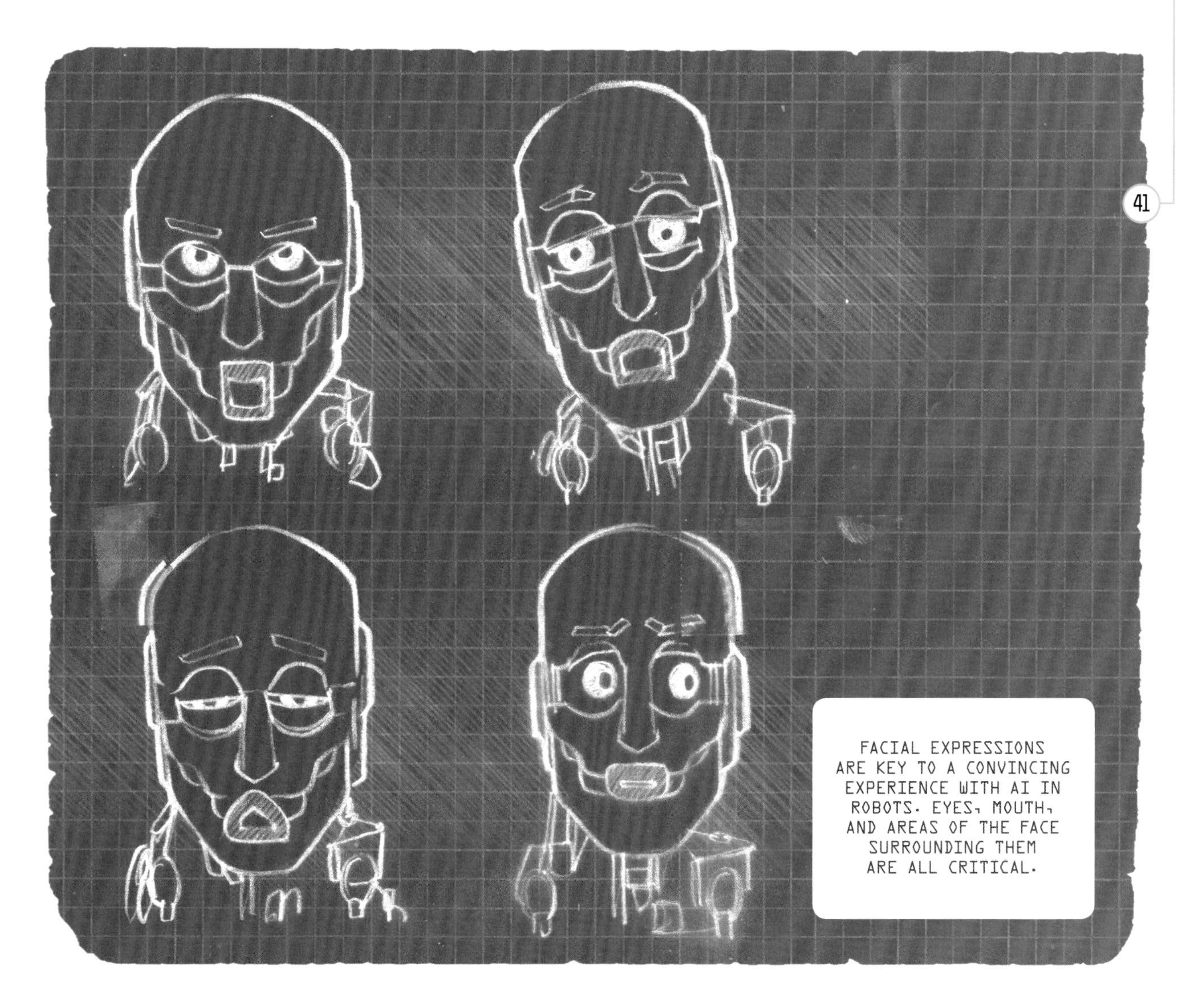

FACIAL EXPRESSIONS ARE KEY TO A CONVINCING EXPERIENCE WITH AI IN ROBOTS. EYES, MOUTH, AND AREAS OF THE FACE SURROUNDING THEM ARE ALL CRITICAL.

STUN, KILL, OR DISINTEGRATE: RAY GUNS AND ROCKET RIFLES

Ray guns are certainly one of the most oft-seen accessories in classic science fiction. Crew members on *Star Trek* were constantly being instructed to set their phaser weapons to kill mode. From the early 20th century onward, anytime a character stepped out of his rocket ship, or traveled into the future, ray guns were an essential part of the kit (whether called laser pistols, phasers, or blasters.) How soon will we be wielding hand held disintegrator pistols?

WEAPONS OF THE FUTURE

In real life, it has been challenging to come up with a ray gun that will accomplish its basic function: zapping an enemy into inactivity, permanent or otherwise. We have had laser pointers for years, but a death ray requires a whole different order of power. Such lasers would have massive energy requirements, suffer interference from the weather, and require your target to stand still while you slowly, ever-so-slowly, burned a hole in his side. And from a tactical standpoint, why bother? Bullet-shooting guns work pretty well at disabling or eliminating threats.

What we really want is Han Solo's blaster. That wasn't a laser, but a 'plasma bolt gun,' and so shouldn't suffer from a laser's limitations (which include the effects of reflective armor diverting much of the laser's energy straight back at its user.) Han's plasma bolt gun looked cool in action, but would have problems of its own.

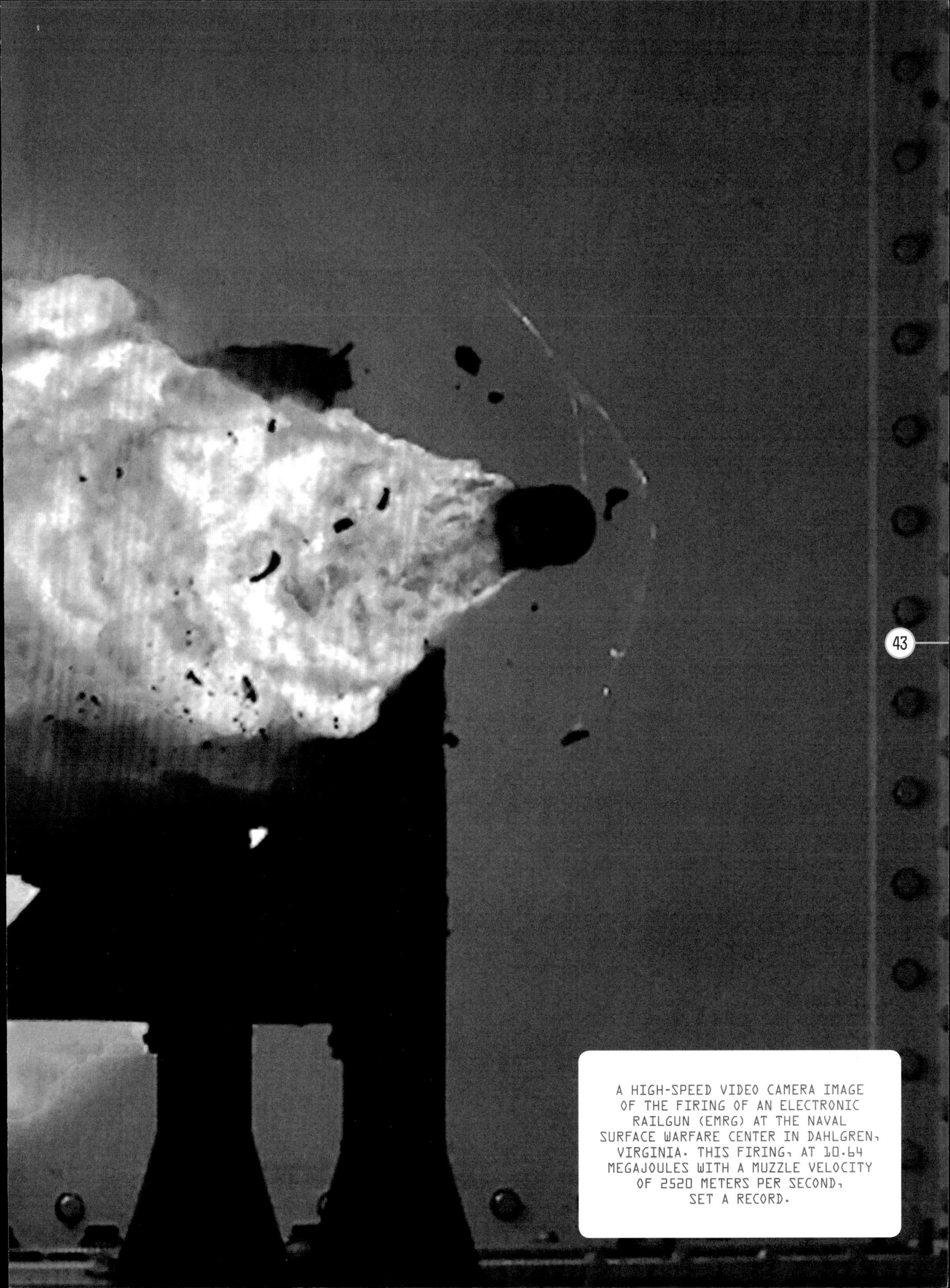

A HIGH-SPEED VIDEO CAMERA IMAGE OF THE FIRING OF AN ELECTRONIC RAILGUN (EMRG) AT THE NAVAL SURFACE WARFARE CENTER IN DAHLGREN, VIRGINIA. THIS FIRING, AT 10.64 MEGAJOULES WITH A MUZZLE VELOCITY OF 2520 METERS PER SECOND, SET A RECORD.

Plasma is simply ionized gas that is able to conduct electrical energy and get really, really hot. It makes a fine torch; it's less spectacular as a light sword. And as a ray gun? Not likely, at least not soon. That blob of hot ionized gas, once it leaves the barrel of the blaster, will do just what any other blob of hot gas wants to do: dissipate and cool. If your target is much farther away than about two feet, it would be useless. Increasing pressure inside the gun and creating a containment field could help, but the power needs would be tremendous.

How about a hand held particle beam weapon? Well, harking back to our discussion of particle beams as death rays (see Chapter 2,) recall that we would need to create a neutral particle beam to prevent it from diverging (due to repulsion) after it leaves the gun. It also takes a lot of power to create a beam of accelerated particles. For many of the same reasons, and greatly amplified by our desire to make this thing a hand held device, particle beam weapons are unlikely to get us what we want in a firearm anytime soon.

In short: getting the destructive, armor-piercing, *Predator*-melting, flesh-burning effect that we want would require dragging an entire power generator station around with you. Discouraging though this

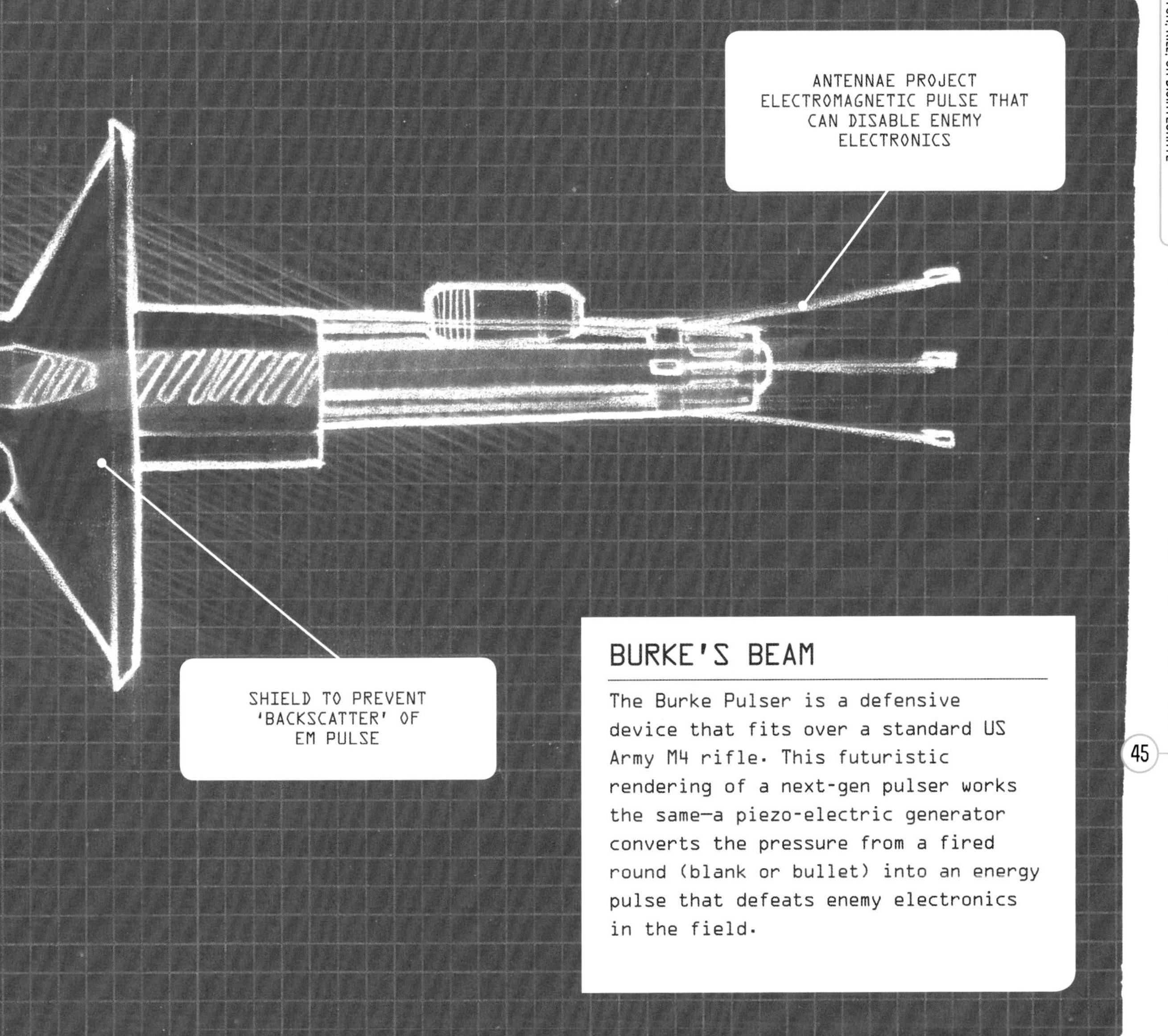

BURKE'S BEAM

The Burke Pulser is a defensive device that fits over a standard US Army M4 rifle. This futuristic rendering of a next-gen pulser works the same—a piezo-electric generator converts the pressure from a fired round (blank or bullet) into an energy pulse that defeats enemy electronics in the field.

might seem, there are real world developments that are bringing versions of a ray gun a little closer.

The shoulder-fired dazzler, despite its rather curious name, is actually an effective and nonlethal anti-personnel and anti-tracking device. These small lasers come in many varieties: some can be held like a rifle, some are mounted on a gun barrel, and still others need a vehicle to carry them. But their basic effect is similar: temporarily blinding your opponent and his electronics (a 1998 UN protocol banned blinding enemy combatants with a laser, but doing so nonpermanently seems to be allowed.)

These dazzlers come in various wavelengths: green lasers are particularly effective against human vision (again, without causing permanent damage if used as directed.) Ultraviolet lasers have also been found to work well as a temporary blinding measure. Infrared lasers are useful to confuse sensors on enemy weapon systems. Most lasers can damage the retina with sufficient exposure, even a few seconds, so devices intended to create temporary effects must be tuned and used with care.

One particularly striking dazzler is the PHaSR (Personnel Halting and Stimulation Response,) a program sponsored by the US Air Force. This

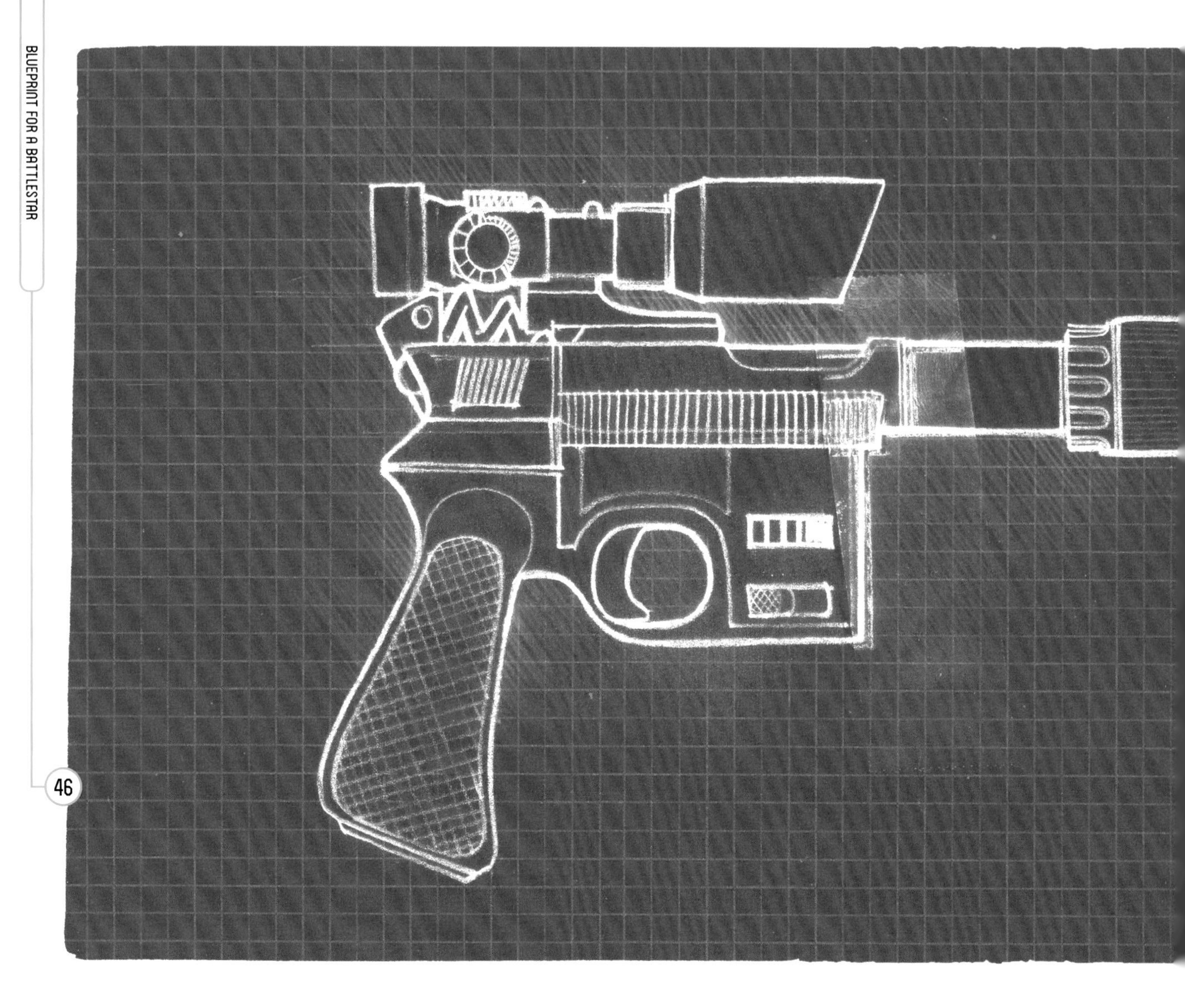

shoulder-fired contraption looks like a massive ray gun, so for appearance, at least, it's a step in the right direction. Just seeing a PHaSR pointed your way would be deterrent enough for most of us.

Functionally, it's pretty ingenious. The rifle operates at sufficiently low wattage to not cause permanent blindness, and uses a distance-ranging laser tool to set the proper intensity. It uses two laser wavelengths, one visible and one in infrared, to cover both visual and electronic threats. Other visual and electronic dazzlers are on the market, but they all work in roughly the same way.

Another nonlethal energy weapon is the Burke Pulser. This innovative little device mounts on the end of a standard military M4 rifle barrel. When the rifle is fired—it can be a bullet or a blank round—the energy from the detonation is captured by a piezoelectric generator, which converts pressure into electrical current. That powers a brief but powerful EMP, or electro-magnetic pulse, which fries any unshielded electronics in the immediate area: drones could fall from the sky, targeting electronics could be damaged, Bluetooth units disabled.

There is money flowing into military laser development, both for units mounted on large trucks, ships, or aircraft, and for portable weapons. Some new technologies are just on the brink of practical use. One uses fiber-optic lasers that bundle multiple laser emitters—usually high-powered LEDs—into one powerful output. The fiber-optic rope can be coiled to give the laser more time to build up power as well (allowing, as we saw

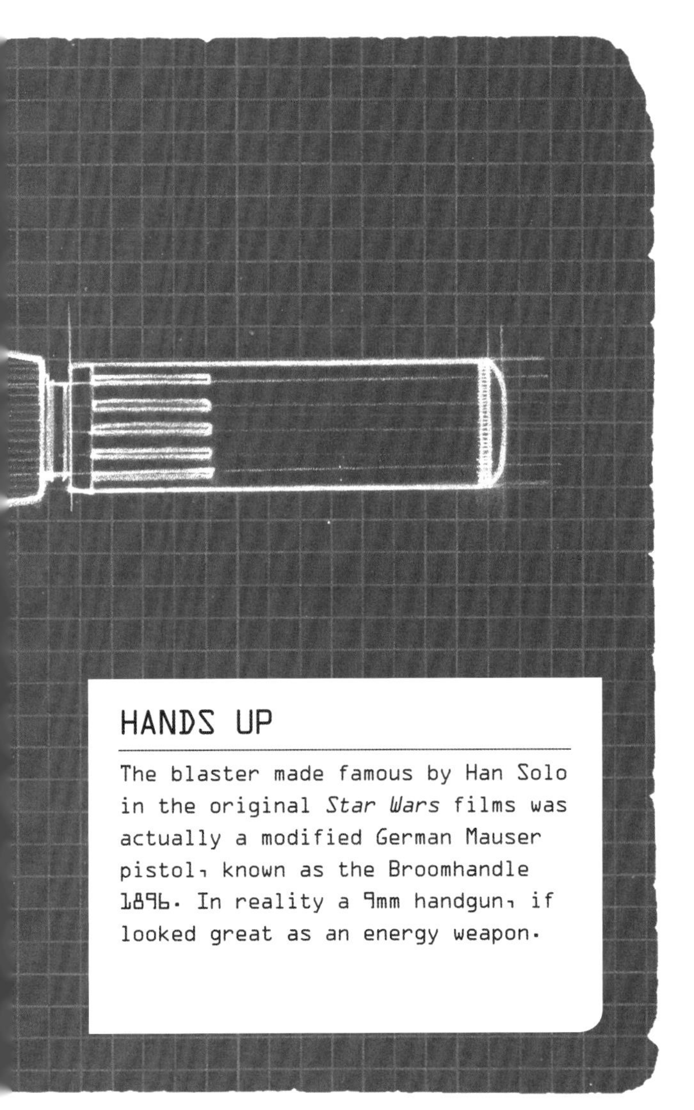

HANDS UP

The blaster made famous by Han Solo in the original *Star Wars* films was actually a modified German Mauser pistol, known as the Broomhandle 1896. In reality a 9mm handgun, if looked great as an energy weapon.

in Chapter 2, the photons to bounce back-and-forth a few times to acquire the energy to exit the laser.) Other blended systems—also using multiple sources to build up more powerful beams in smaller space—are soon th be deployed.

There is also some ingenious engineering by amateurs going on. One bright young man in the US homebuilt his own shoulder-fired railgun, the same kind of atom-smasher derived device that the US Navy has been experimenting with (on a vastly larger scale) for years. The young inventor created much of his railgun on a 3-D printer and it uses over 20 lb of electrical capacitors to fire aluminum slugs at almost the same velocity as a .22 caliber rifle.

Then there is the rocket rifle. In the 1960s, a couple of gentlemen apparently decided that while the army's bazooka shoulder-fired rocket launcher was pretty neat, why not make a hand held version? The result of their fascination was, after much trial and error, the Gyrojet rocket gun.

What set the Gyrojet apart from other firearms was that it was not really a firearm; it was a tube with a triggering device. The metal bullets had small plugs of solid rocket fuel in them, with four tiny jet-exhaust ports machined into the back of the projectile at an angle. When the gun was fired, there was no 'bang' and no recoil, just a 'whoosh' as the rocket ignited. The bullet left the barrel, spinning furiously for stability.

Unfortunately, the exit velocity was slow and it took over 50 feet for the projectile to pick up speed, though by that time it was traveling at roughly 250 feet per second, and imparted much more destructive force than a traditional .45 caliber bullet. The army reportedly once tested it by firing a round *through* a tree trunk. The gun did have its advantages: besides the additional punch of the round, it could also be fired in space or even underwater.

The Gyrojet made an appearance in the 1967 Bond movie *You Only Live Twice* as the head of Japanese intelligence is showing Bond some of their slick new gadgets. It was the Gyrojet's finest hour. Ultimately, the gun was produced in both pistol and rifle versions. It had some potential, but the rounds proved to be unreliable, and not sufficiently powerful at short range. Accuracy was also a problem, and when the gun fired, blasts of rocket exhaust were vented out the sides of the slide, temporarily blinding the user. Finally, as with so many innovations, it was a solution to a problem nobody had. We weren't fighting wars in space or underwater; this was the Vietnam era and plain old bullets were much in vogue. The Gyrojet died a quiet death as a result of disinterest.

Hand held and shoulder-fired energy weapons capable of creating mayhem and destruction are still in the future, but the output of smaller, solid-state lasers is increasing in a way somewhat like the Moore's Law curve of microprocessors, though more slowly. For better or worse, laser death rays will happen.

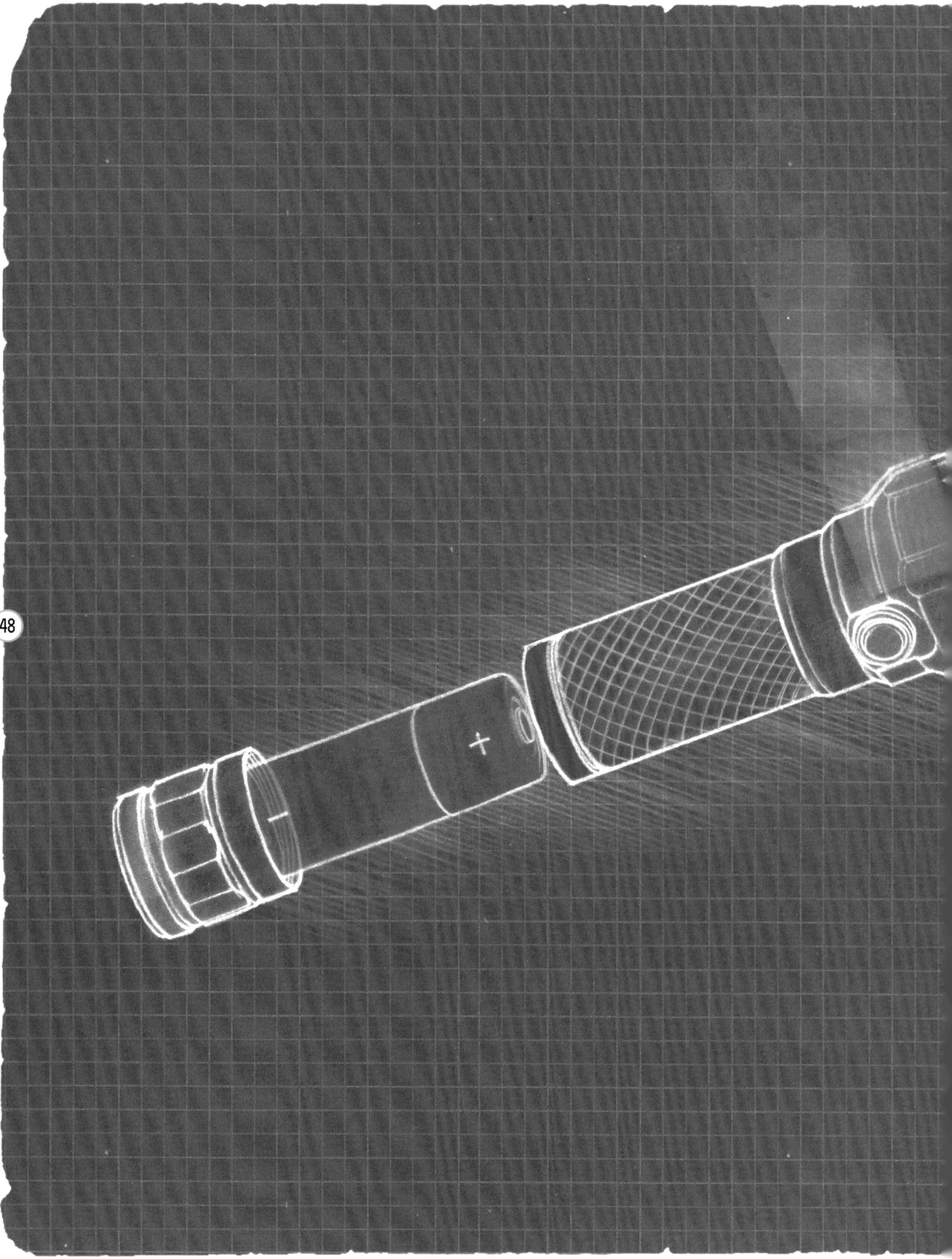

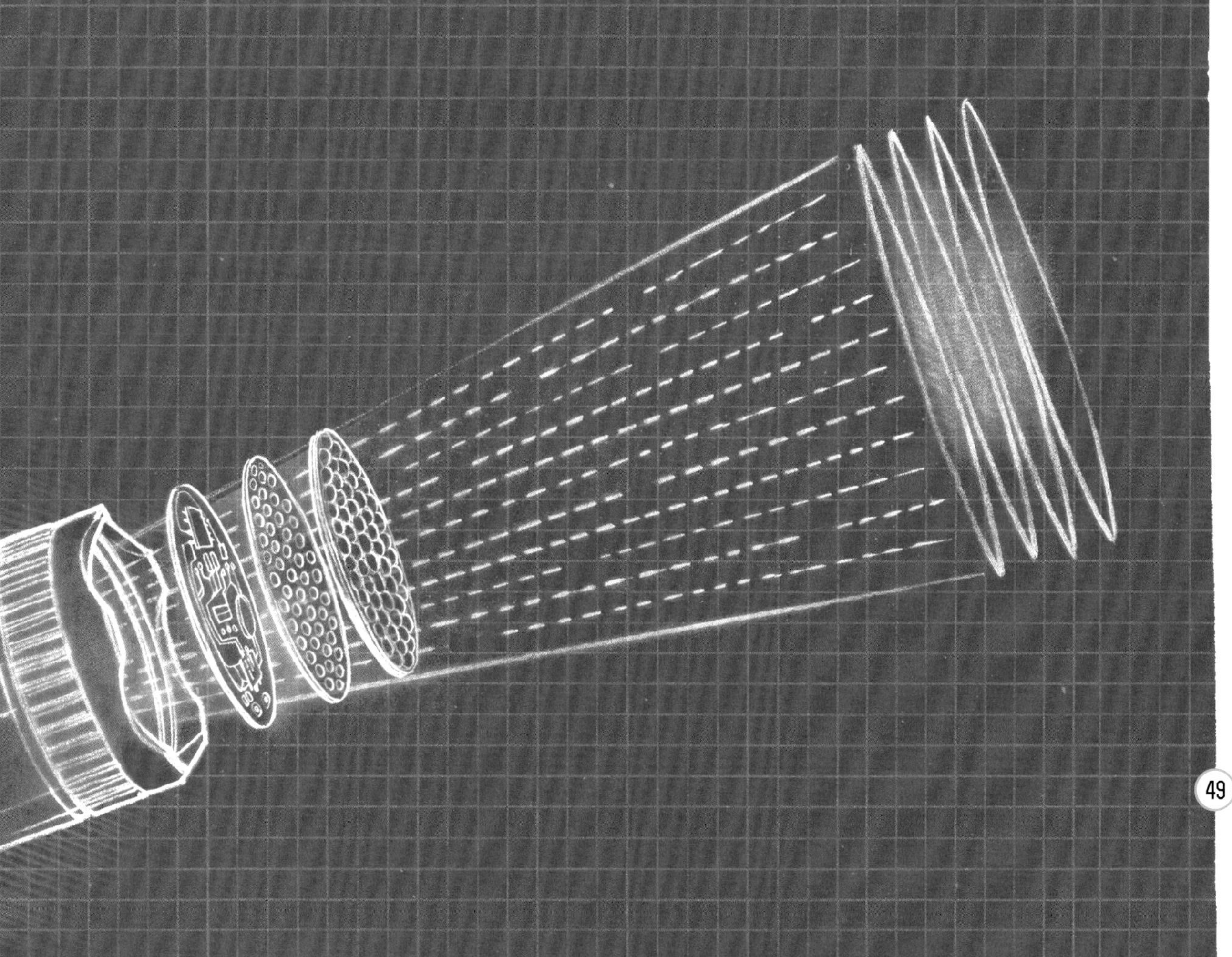

THE INCAPACITATOR

This seemingly innocent flashlight will cause you to puke before you know it by using high-intensity LEDs rotating across a series of rapidly changing colors. As the brain tries to focus one on color, it changes in a bright, strobing fashion, causing intercranial pressure, headaches, and intense nausea. It's not called the barf-beam for nothing.

FANTASTIC VOYAGES

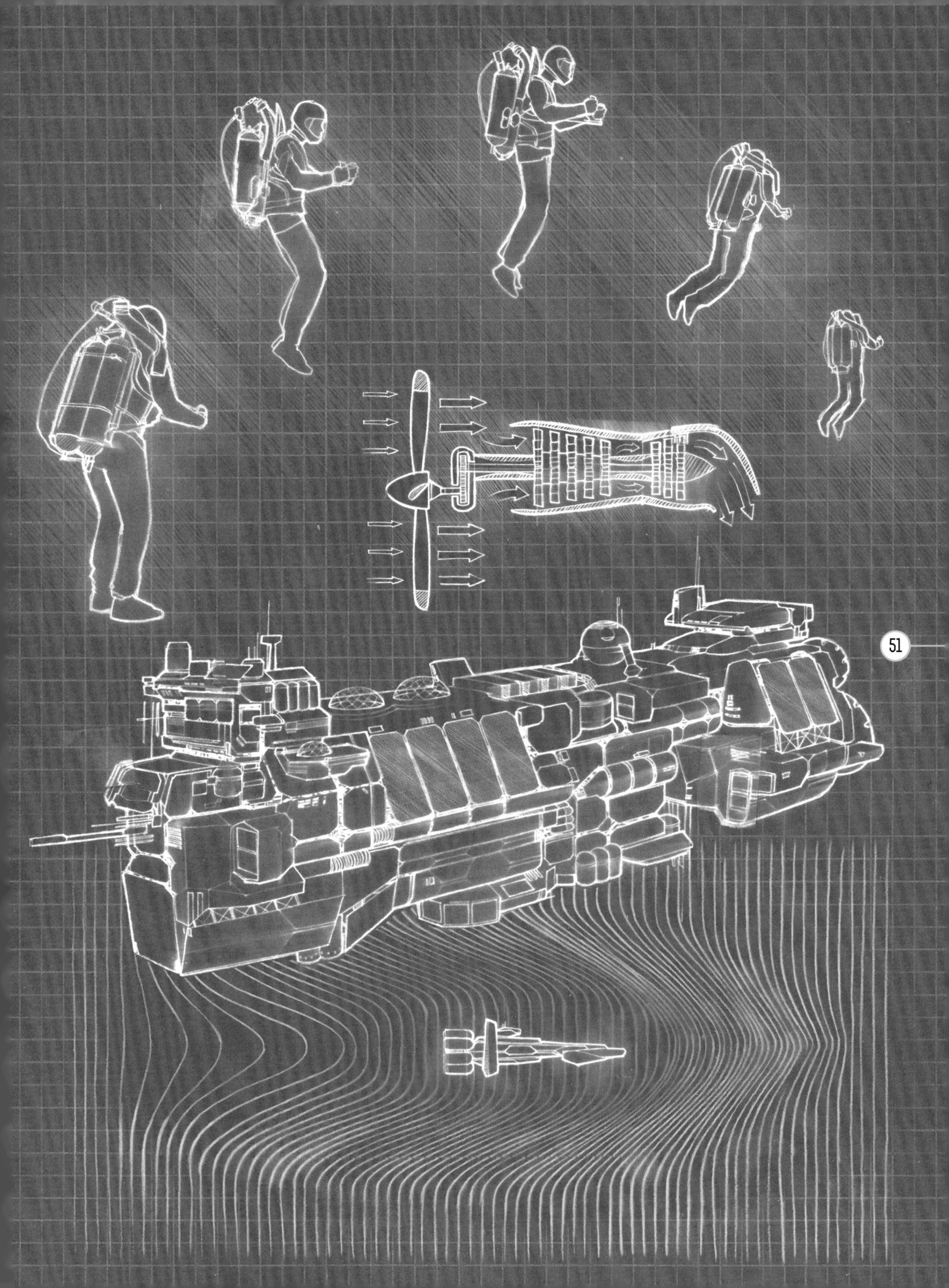

ROCKET SCIENCE: HOW TO FLY A JET PACK

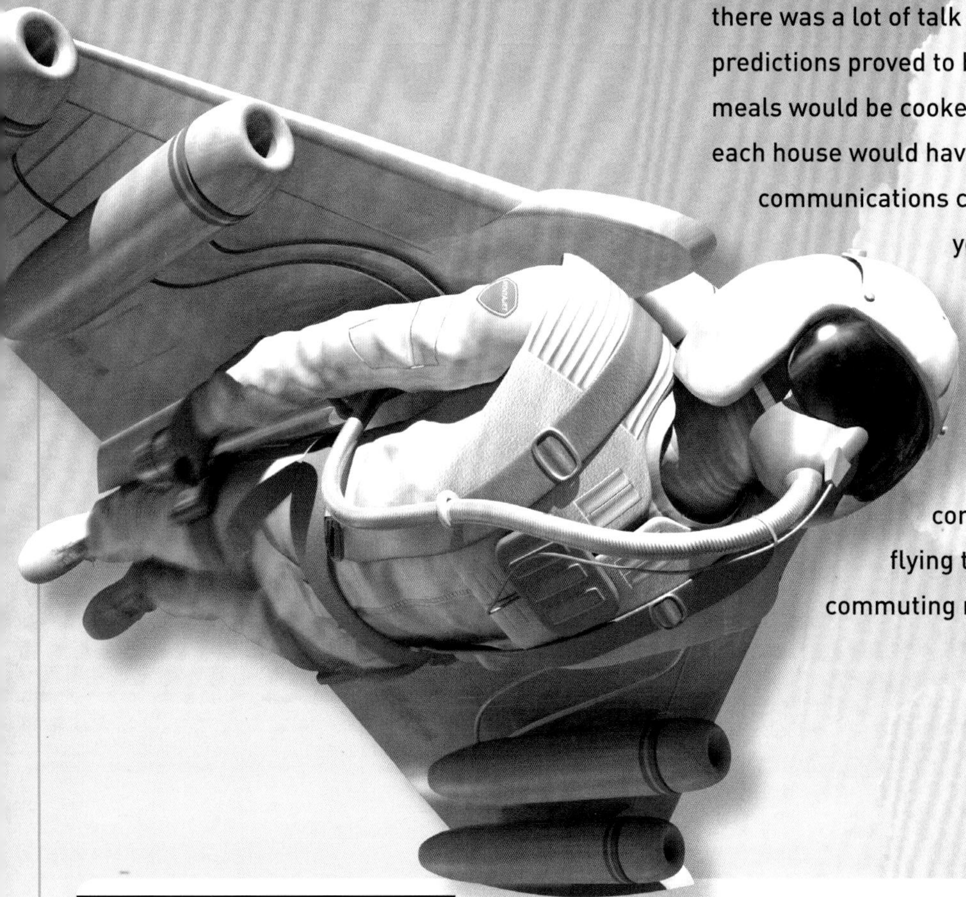

In the 1960s, *Popular Science* magazine was a mainstay in many homes. In 1966, at the dizzying height of the space race, when anything seemed possible if enough technology was thrown at it, there was a lot of talk about the future. Some of the predictions proved to be true or nearly true—that meals would be cooked with radio waves and that each house would have a computerized communications center. Others, sadly, have not yet come to pass, such as the forecast of vacations on the moon. One in particular caught the imagination: the Bell Aerosystems Rocket Belt. The magazines confidently predicted a future of flying to work and an end to commuting nightmares.

A DREAM OF THE AGES

Rocket belts, alternately known as rocket packs or jet packs, were staple fare in science fiction from the 1920s on. Buck Rogers, hero of the 25th century, flew one and in real life there were (unsubstantiated) rumors that Nazi Germany was trying to develop a jet pack during World War II. Even James Bond flew one in the 1965 movie *Thunderball*.

But jet packs never arrived. The most promising of the early breed, Bell's Rocket Belt, had a flight time under 30 seconds, and was dangerous to operate, requiring a highly skilled operator. The military, though, expressed great interest; the idea of soldiers being able to fly over minefields, crossing great distances in a flash and descending on an

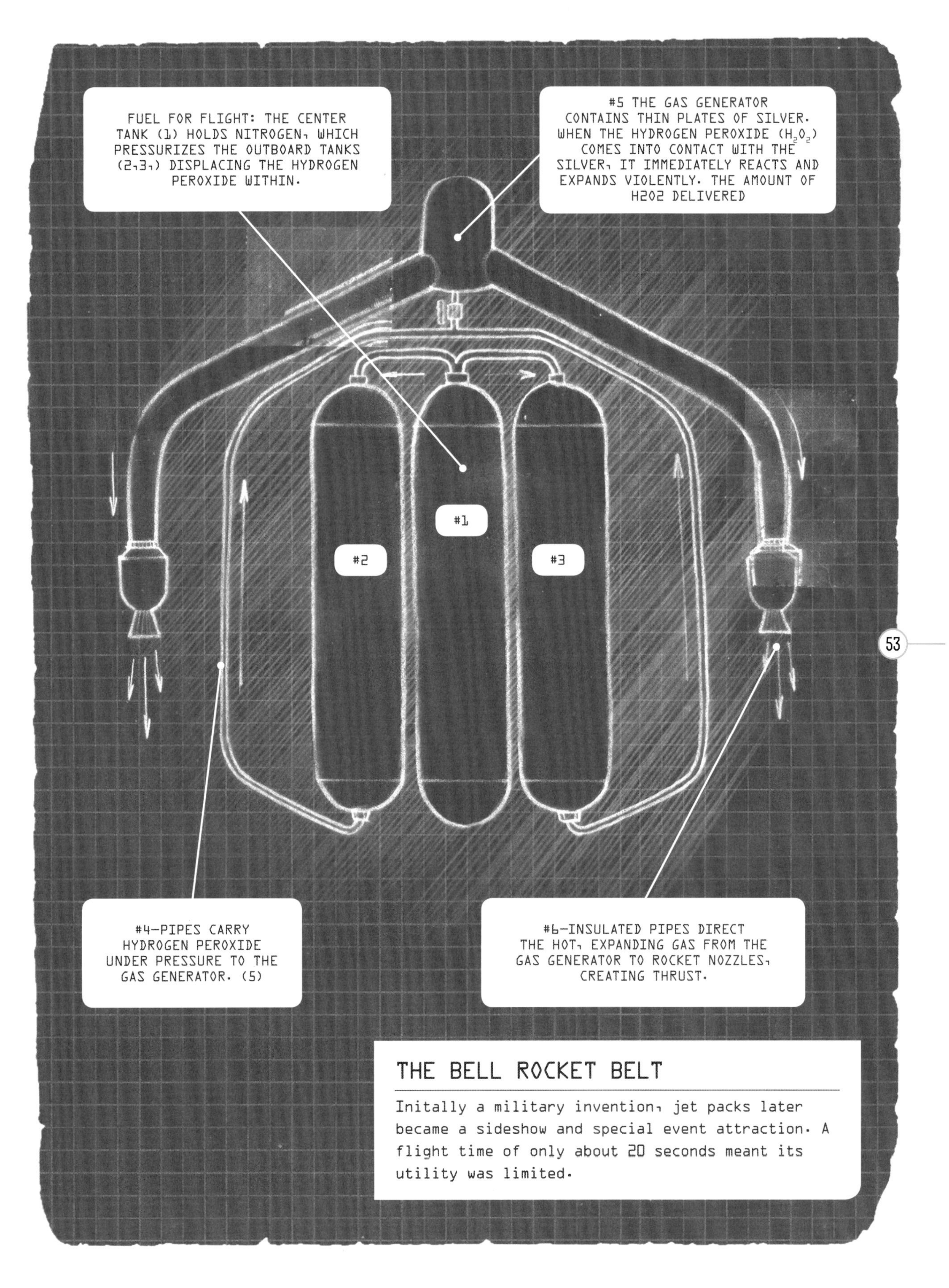

THE BELL ROCKET BELT

Initally a military invention, jet packs later became a sideshow and special event attraction. A flight time of only about 20 seconds meant its utility was limited.

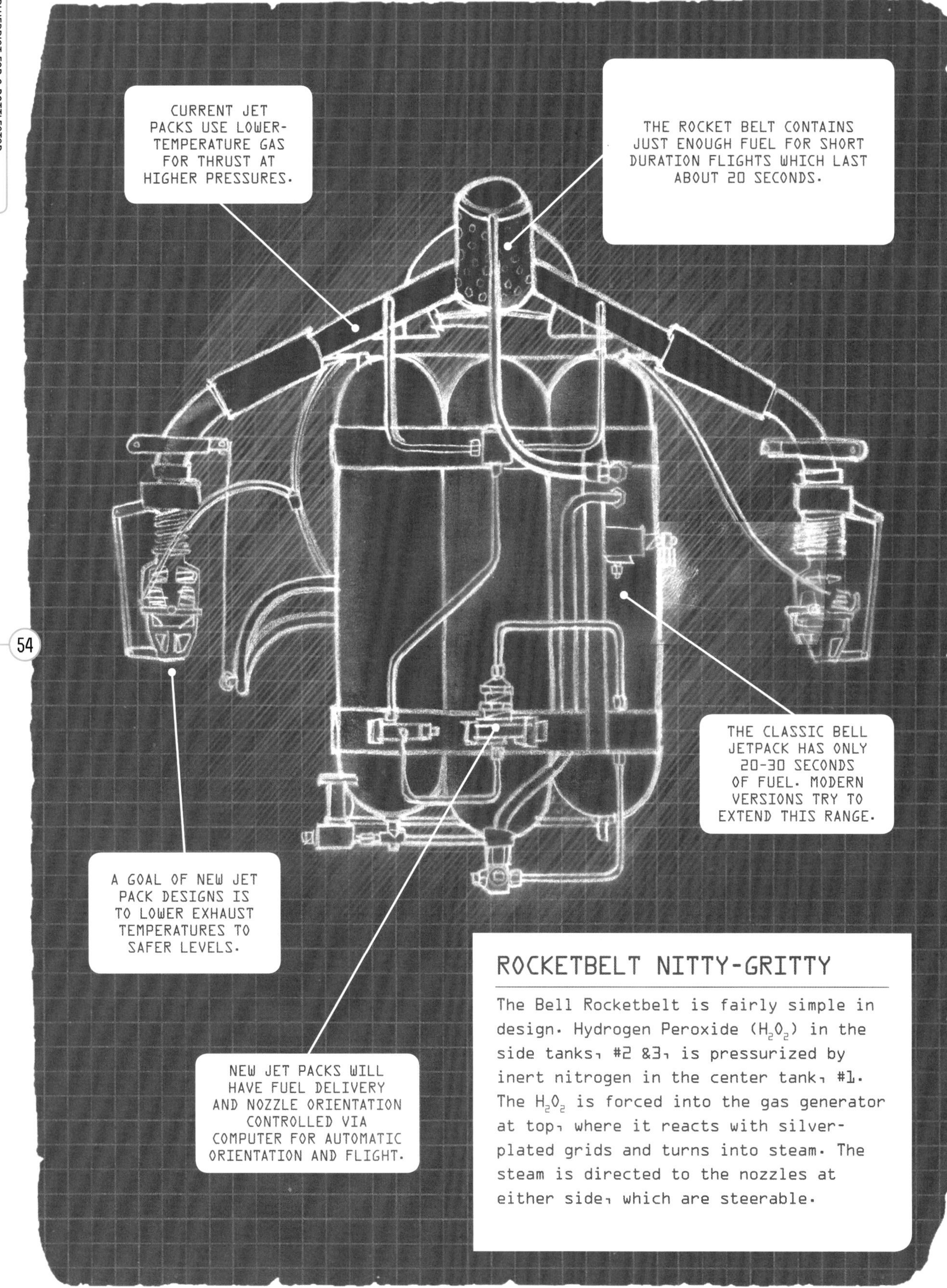

ROCKETBELT NITTY-GRITTY

The Bell Rocketbelt is fairly simple in design. Hydrogen Peroxide (H_2O_2) in the side tanks, #2 &3, is pressurized by inert nitrogen in the center tank, #1. The H_2O_2 is forced into the gas generator at top, where it reacts with silver-plated grids and turns into steam. The steam is directed to the nozzles at either side, which are steerable.

unsuspecting enemy, was just too good to pass up. But the short flight time, complexity of operation, and the skull-splitting 130-decibel noise (equivalent to a jet taking off a few hundred feet away) made rocket-powered commandoes impractical. The US Army lost interest, and the last space age application was a NASA-built design that was sent into space, but never used, on the Gemini 9 spaceflight.

The design of the Bell unit and its successor rocket belts is straightforward. Hydrogen peroxide was used as the fuel. This was pressurized by small amounts of liquid nitrogen, which is inert. The pressurized hydrogen peroxide flowed over a screen of silver, which catalyzed it and explosively released steam. The steam was directed out of two downward-facing nozzles and provided the thrust to make the 'rocketman' fly, albeit for just a few seconds. Mechanically simple, it proved complex in operation. The pilot had to be highly trained, and the concentrated fuel was expensive (at over $250 per gallon). The pilot faced additional hazards, with exhaust plumes of over 700°F and an operating altitude below that at which a parachute would be of any assistance.

The main limitation of this early jet pack was that carrying both the fuel and oxidizer that it needed to function meant it was heavy, and the downward thrust necessary to overcome gravity was considerable. Even so, Bell's successor company worked with NASA in an attempt to design a mobility system for use by American astronauts on the moon, but in the end the space agency wisely decided to send their astronauts out across the lunar surface with a car-like moon rover.

A $125,000 model is still produced by a Mexican company based on the Bell design, but the Bell Rocket Belt essentially remains a novelty, occasionally turning up at events such as the famed Rose Parade in Pasadena, California, in 2007. Variants of it, though, did see use in outer space. During a 1984 space shuttle mission, astronaut Bruce McCandless strapped-on the Manned Maneuvering Unit or MMU, and became the first free-flying human in space. It was a large and complex affair, but was used successfully on three shuttle flights. Even so the risk of an accident (and a fatal one-man re-entry to Earth) was deemed to be too high and it was quietly shelved. Only a smaller version, the Simplified Aid for EVA Rescue (or SAFER), survived as a means to rescue a spacewalking astronaut who has drifted free from the tether.

A far more effective way than explosive thrusters to power a rocket pack is by using jets. Swiss pilot Yves Rossy demonstrated a turbofan jet pack in the early 2000s against stunning mountainous backdrops, using small jet engines fixed to the bottom of a carbon fiber composite wing that could travel in excess of 15 minutes at a top speed of 180 mph. In fact, the Bell company had already experimented with ducted fan jets in the 1960s, but the project hit a dead end. Rossy's higher profile efforts performed spectacularly and enthralled crowds with the possibilities of a future featuring personal jet packs.

Further technical developments promise even more spectacular jet packs. A company called JetPack Aviation came to the fore with a public exhibition of their JB-9 jet pack in 2015, featuring a dramatic loop around the Statue of Liberty. It is easy to operate: the user simply starts it up in a standing position and launches for up to ten minutes of exhilarating free-flight. It looks a lot like the old Bell rocket belt, except that two fire extinguisher-sized jet engines are mounted where the Bell's exhaust nozzles would be. It also runs off of cheap, readily available kerosene and uses air as an oxidizer.

This may be as close as we get to the original idea of a real jet pack unless someone invents an antigravity pellet or magneto-levitation. The JB-9 fits in a large suitcase and can be worn by an average person without difficulty. The jet exhaust is said to be warm, but not scalding, and the engines can be tilted via the hand controls, offering superb maneuverability. At 125 mph, it is a bit slower than Rossy's personal jet-wing, but does not require jumping out of an airplane to get started. The climb rate is a bracing 1,000 feet per minute and it currently has a 10,000-foot ceiling. The company is working on an electronic package to enhance safety, which would enable users to rise from the ground and let the JB-9 (or 10, or 11 by that time) hover unassisted. They are even working on a jet pack racing series, where trained pilots will don the units and fly around various checkpoints in a race to the finish (hopefully, a metaphorical one.)

Pricewise, the JB will remain out of reach of the average person, but budget jet packing can still be experienced in the form of the modified Jet-Ski. The Flyboard, manufactured by Zapata Racing, allows jet pack-like aerobatics for the rest of us. The original model looks like a skateboard with a couple of nozzles behind the shoulders, positioned like the Bell Rocket Belt. The user wades out into shallow water, where their feet are strapped into what look like ski boots mounted on a transverse board. The nozzles are mounted to shoulder-height pivots that are controlled via handgrips near your waist. When a nearby Jet-Ski is fired up, instead of firing a torrent of water out the back end to propel itself through the water, it pumps copious amounts of water through what looks like a fire hose that is attached to the Flyboard. The water gushes from the nozzles, forcing the Flyboard upwards. The genius of the Flyboard is that all the propulsive mass—energy source, pumps, and fuel or reaction mass stays on the ground (or, in this case, water.) The Flyboard is, of course, limited to the length of the hose, but with a range of about 60 feet, it still provides the sense of jet pack flying at a fraction the cost of the more technically sophisticated alternatives.

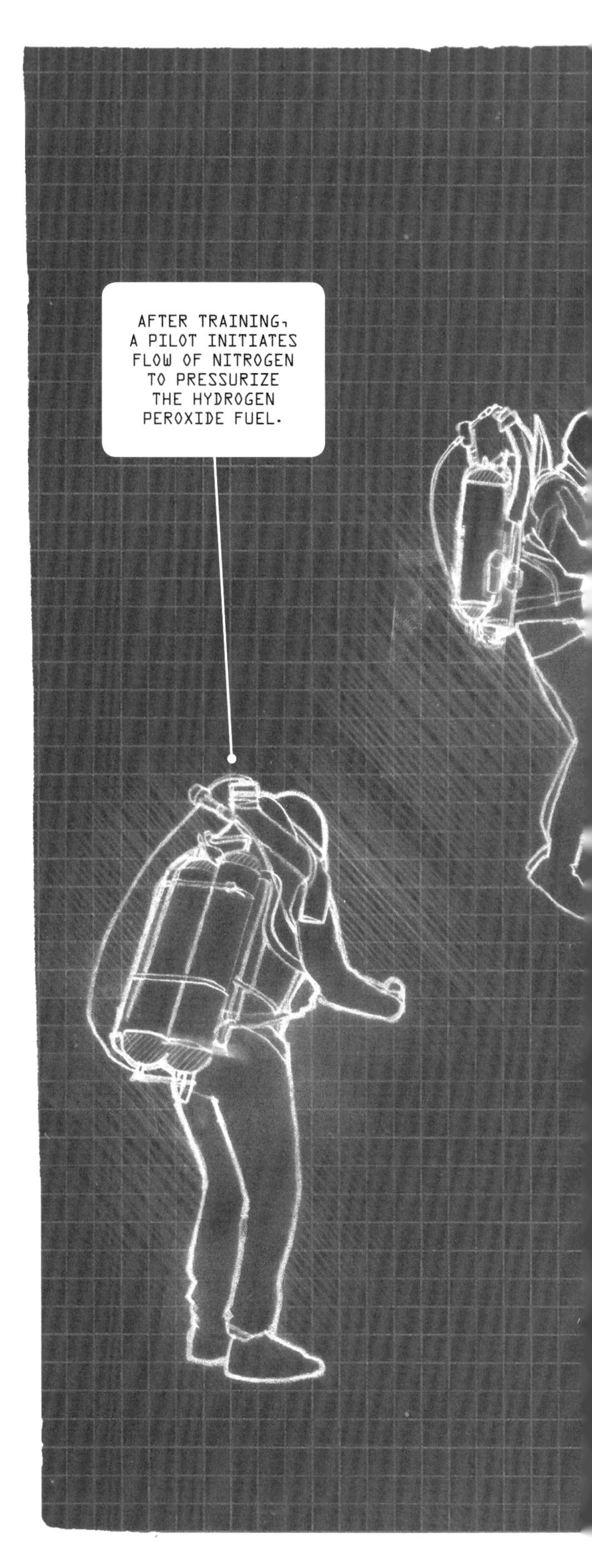

MAX ALTITUDE: APPROX 60 FEET
FLIGHT TIME: APPROX 20 SECONDS
MAX SPEED: 30 MPH

CAUTION HAD TO BE USED WHEN MANEUVERING—THE JET PACK FLEW AT TOO LOW AN ALTITUDE FOR A PARACHUTE TO BE EFFECTIVE.

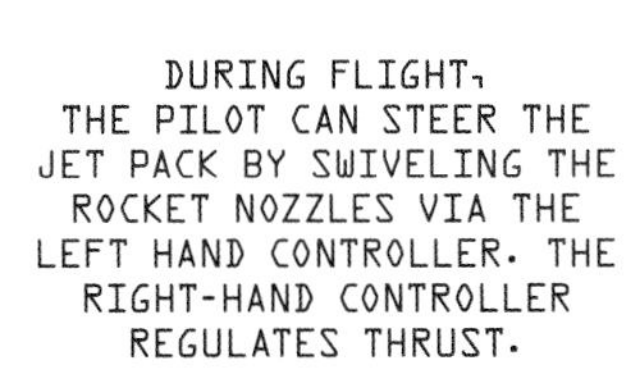

LANDING WAS ACCOMPLISHED BY REDUCING THRUST AND SETTING-DOWN WITH KNEES BENT TO ABSORB IMPACT, MUCH AS ONE LANDS USING A PARACHUTE.

PHASES OF FLIGHT

Most jet packs are have been based on the old Bell Aerosystems Rocket Belt. Despite improvements, it is still a specialty attraction.

WORMHOLES: TRAVELING FASTER-THAN-LIGHT

Many written science fiction stories, and most televised ones, take the idea of being able to 'warp' a few light years at a moment's notice as a given. But is traveling at or faster than the speed of light actually possible? Or perhaps it would be good enough just to travel some appreciable percentage of the speed of light (commonly abbreviated as *c*)? Light travels at 186,282 miles per *second*. That's fast. And even at that speed, it would take light over four years to reach the nearest star.

EINSTEIN'S THEORY

Albert Einstein discussed the limits of such travel in his Special Theory of Relativity, published in 1906. His final paper on General Relativity followed in 1915. These ideas became the conceptual lens through which we view the universe; with a few challenges, they have stood the test of time well.

Einstein showed that an object in motion, as it accelerates, will experience an increase in mass, a contraction in the direction of travel, and a dilation of time (time slows down as viewed by a stationary observer.) So, for example, a spacecraft traveling at about ninety percent of light-speed, or *c*, would: increase in mass by over two times, appear to contract by over half its length, and, if you were onboard, 'your' year would be 2.2 years as experienced back on Earth. Confusing, isn't it?

Extrapolating these effects to ever-faster velocities, you can see that going at *c* or faster than *c* gets tricky, if it is possible at all. As you approach c, mass and contraction go way off the scale toward infinity. Einstein claimed that the effect was relative to a static observer, and that the passenger in our imaginary spaceship would not notice it...but this has yet to be tried.) It also appears to mean that ultra-long-distance travel cannot be accomplished by simply going fast.

Speaking of speed, how fast have we traveled to date? During the Apollo lunar landing program of the 1960s, the Apollo 10 spacecraft reached a top speed of 24,791 mph as it sped back to Earth from the Moon, which is still the fastest that humans have traveled. Yet, while it took Apollo three days to reach the moon; light takes just over a second to make the same journey. Apollo 10 traveled at about 0.0037 percent of c, so the Saturn V moon rocket that powered it won't be taking us to the stars. In fact, chemical rockets—the kind that powered Apollo, the space shuttle and all other manned space endeavors to date—will not be of much help at all. They are simply not powerful enough and have to drag too much heavy fuel around.

Unmanned space probes do travel faster. A robotic spacecraft called Helios 2, was launched in 1976 to orbit and study the sun, and sped along at about 150,000 m.p.h. as it swung past our star, the largest gravitational force in the solar system. That's about 0.023 percent of the speed of light. Bear in mind, these speeds are in miles per *hour*, whereas light-speed is usually quoted in miles per *second*. If you want to use the standard measurements, light-speed is more like 670,616,629 m.p.h.

Clearly, making interstellar voyages in the kind of spacecraft we have so far been using is impractical. So what kinds of technology might help us along the path to *Star Wars*-style FTL drives? There have been efforts to design faster spacecraft in the past. Two of the more courageous plans are mentioned here.

In the late 1950s, an audacious program called Project Orion was initiated, championed by the renowned physicist Freeman Dyson, among others. The general idea was to build a giant, crewed interplanetary spacecraft—one capable of going to Mars, Jupiter, and beyond, powered by nuclear detonations (atom bombs for smaller designs, hydrogen bombs for the big, interstellar versions.)

The spacecraft would have a crew and cargo area at the top, a fuel storage area in the middle (containing thousands of tiny atom bombs that slid down tubes to the tail end,) and a giant 'pusher-plate' on the bottom with holes through which the bombs would pass before detonating. This would absorb the bomb blast and propel the giant spacecraft, with shock absorbers to take up some of the punishing acceleration. The total mass of the spaceship, depending on the design iteration, ranged from 100,000 tons to 10 *million* tons. It was far easier to design on paper than to build since it would have to be at least 5000 tons to work, double the mass of the Saturn V rocket. We will hear more about Orion in the Chapter 9, but the important thing here is that in its fastest variant, with thousands of bombs aboard, it would eventually travel at 3.3 percent of c and would be capable of reaching Alpha Centauri (the nearest star system to our own at just 4.4 light years away) in 'only' 133 years (with additional time required to slow down, if you planned to stop there for a visit). It was a wild but, theoretically, feasible idea, and General Atomics, the firm that first championed the idea, worked on it for a decade before interest and funding dried up.

Test flights of a miniature version, with a diameter of three feet and using conventional explosives, actually resulted in the machine being propelled into the air to a height of about 150 feet. That was about as far as they got, however. The idea of using at least 800 nuclear explosions just to get the full-sized craft into orbit, which statistically would kill between one and ten people per launch from the effects of nuclear fallout, was a drawback, to say the least; as was the cost of between triple-digit billions to trillions of 1968 dollars The entire Apollo program cost less than $20 billion.

Another fascinating plan was called the Bussard Ramjet, sometimes referred to as a cosmic ramjet. This represented another enormous engineering project. The ship proposed by physicist R.W. Bussard in 1960 would be 1,000 tons, far less massive than Orion, and was to be powered by nuclear fusion. At the front was an enormous funnel-shaped antenna made of wire mesh that would generate an electromagnetic field about 60 miles in diameter, which would also be cone-shaped. As the craft plowed through space, this field, acting like an enormous scoop, would gobble up hydrogen atoms from the interstellar void (the universe is made primarily of hydrogen, so even in the 'empty' space between the stars, it was thought at the time that there would be a lot of it). This would be funneled into the body of the spacecraft, where the fusion reactor would contain the extremely powerful fusion reactions between hydrogen nuclei, with the resulting energetic reaction directed backward to provide a propulsive exhaust.

Due to the high energy released by nuclear fusion, and the fact that the spacecraft literally fuels itself, Bussard estimated that the craft would be able to reach nearly the speed of light after traveling for a year and would be capable of going right to 'the edge of the universe.' Later (and more realistic) estimates suggest that it would probably not top 30 percent of *c*. Other theoreticians have postulated that the design cannot work at all, explaining that there is apparently less hydrogen in most of space than once thought or pointing out that deuterium—known as 'heavy hydrogen,' with an extra proton in the nucleus—is a far better fuel source for fusion than 'regular' hydrogen.

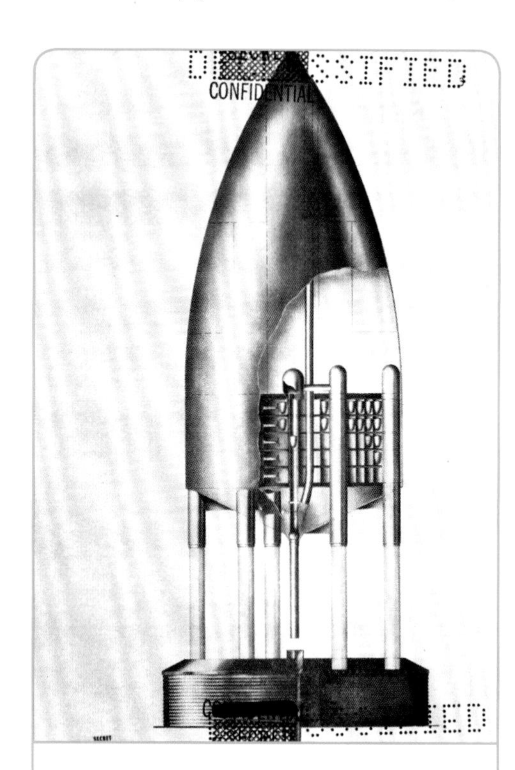

Project Orion

Project Orion was a concept for an interplanetary (and possible interstellar) spacecraft and was extensively studied in the 1950s and 1960s. It was massive and would be propelled by nuclear bombs shoved out the back and detonated, creating thrust. Other than non-nuclear explosive propulsion testing, it never made it off the drawing board, but might actually have worked. Orion would have flown faster than anything since put into space.

All of these design ideas are either sub-light speed or (in Bussard's dreams) barely approaching *c*. The only way to get around Einstein's relativistic bar on faster-than-light travel seems to be to sidestep traditional physics altogether. It is no surprise that most science fiction faster-than-light travel has involved such methods, either by twisting or 'warping' space (as used in *Star Trek* and *Star Wars*) or to take pre-made (or custom-formed) shortcuts via 'wormholes.'

Let's look at wormholes first. In a form that became known as black holes, these have been speculated about since shortly after General Relativity was published. This was refined by Einstein in 1935 and called an Einstein-Rosen Bridge (what we now call a wormhole—a completely theoretical construct.) But these are massively powerful things—the gravitational effects could really mess up your trip—compressing you and giving you infinite mass, not to mention the endless amounts of radiation you'd be fried to burnt atoms by.

So for a long time, it looked like wormholes were an amusing abstraction, but not ultimately useful for future space travel...until 1988. Then, Kip Thorne, a professor at the California Institute of Technology, proposed a 'traversable' black hole, which might insulate objects passing through it sufficiently that a human could survive. Of course, you still need to find or make one before you can use it.

If we did find a wormhole that has occurred naturally (or been created by some advanced technology,) we still have to get to its location, which could turn out to be relatively nearby or very far away. To create one, we would need enormous energy. This kind of portal also appears to require a type of

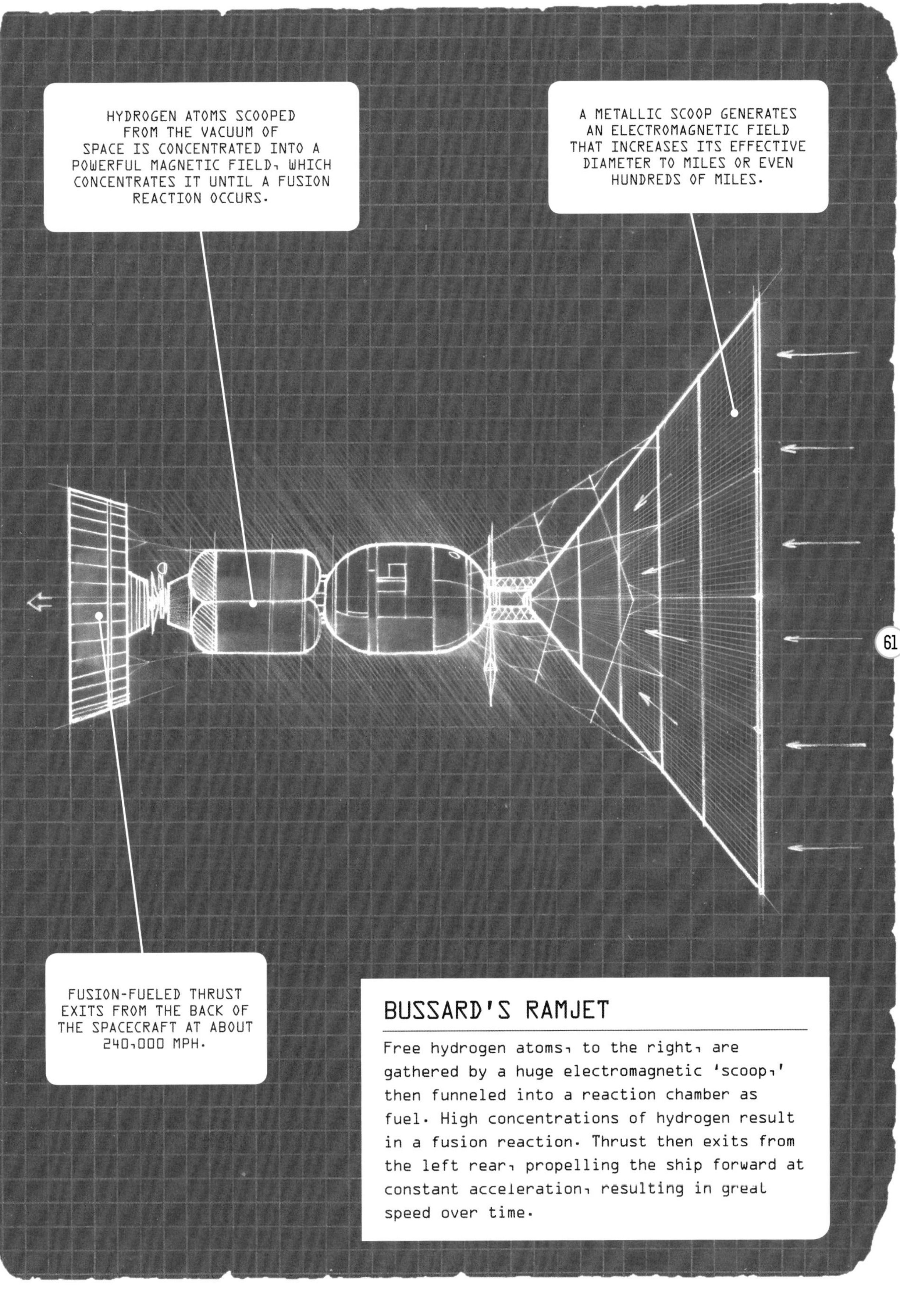

HYDROGEN ATOMS SCOOPED FROM THE VACUUM OF SPACE IS CONCENTRATED INTO A POWERFUL MAGNETIC FIELD, WHICH CONCENTRATES IT UNTIL A FUSION REACTION OCCURS.

A METALLIC SCOOP GENERATES AN ELECTROMAGNETIC FIELD THAT INCREASES ITS EFFECTIVE DIAMETER TO MILES OR EVEN HUNDREDS OF MILES.

FUSION-FUELED THRUST EXITS FROM THE BACK OF THE SPACECRAFT AT ABOUT 240,000 MPH.

BUSSARD'S RAMJET

Free hydrogen atoms, to the right, are gathered by a huge electromagnetic 'scoop,' then funneled into a reaction chamber as fuel. High concentrations of hydrogen result in a fusion reaction. Thrust then exits from the left rear, propelling the ship forward at constant acceleration, resulting in great speed over time.

exotic matter called negative energy; in effect, less-than-nothing in space. Negative energy has been created in the laboratory, in minute amounts. But for our wormhole, one physicist estimates that to make a three-foot wide wormhole you would need a mass of negative energy equivalent to that of Jupiter. It's hardly a practical option.

In a sci-fi setting such as *Star Trek*, the warping of space is accomplished via machines. Matter-antimatter reactions provide the 'warp' around the *Enterprise*, and the deflector dish keeps the route ahead clear of anything the ship might run in to. If you warped space properly, you could compress it in one direction and expand it in another, the effect of which would be, theoretically, to enable you to travel across vast distances in short order. And since you are in a warp-bubble, you are not experiencing the dynamics of acceleration (which would turn you to a red smear), or time-dilation (which would freeze time for you, but when you got home, everyone you knew would be long dead).

Most physicists used to say smugly that this was all amusing fiction. Then, in 1994, a young Mexican physicist named Miguel Alcubierre came up with the notion that it would be possible to travel faster-than-light under certain conditions that circumvent the usual limitations of relativity. His mathematical work essentially takes general relativity, bags it up, and takes it with you in the spaceship. To be a bit more formal, his model enables a bit of 'normal' or flat space—big enough for a spaceship—to be transported inside a 'bubble' of curved space. The bubble, which he called *hyper-relativistic local-dynamic space*, is propelled forward by the local expansion of space behind it and an opposite contraction in front of it, a bit like the way a maglev high-speed train is propelled with magnets. Or, if you like, surfing on a wave-front in warped space. So a ship using what has come to be called the 'Alcubierre Drive' is placed in motion by changes made in the realm of space-time and travels within a distortion bubble that keeps things inside normal; there is no traveling faster than *c inside* the bubble, so it doesn't violate the older, accepted rules that apply to the outside world.

There have been, of course, people who doubted the validity of this idea. They point out that just because something like Alcubierre's drive can be made to work

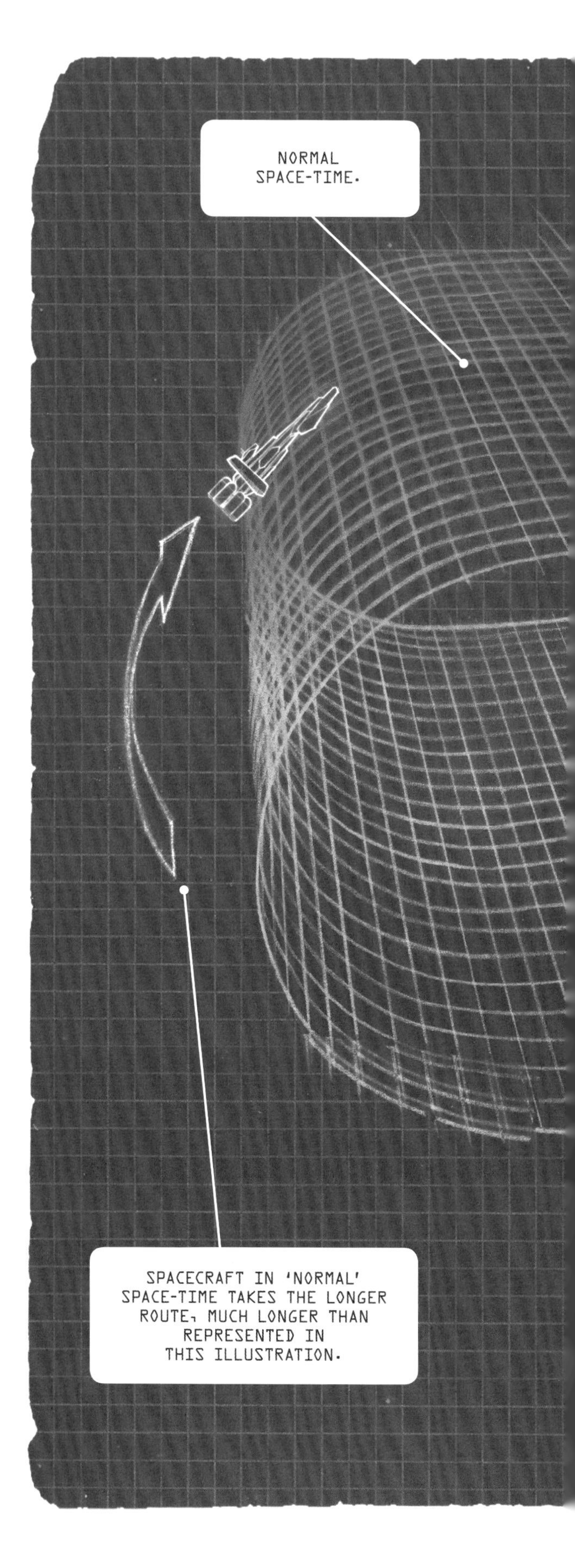

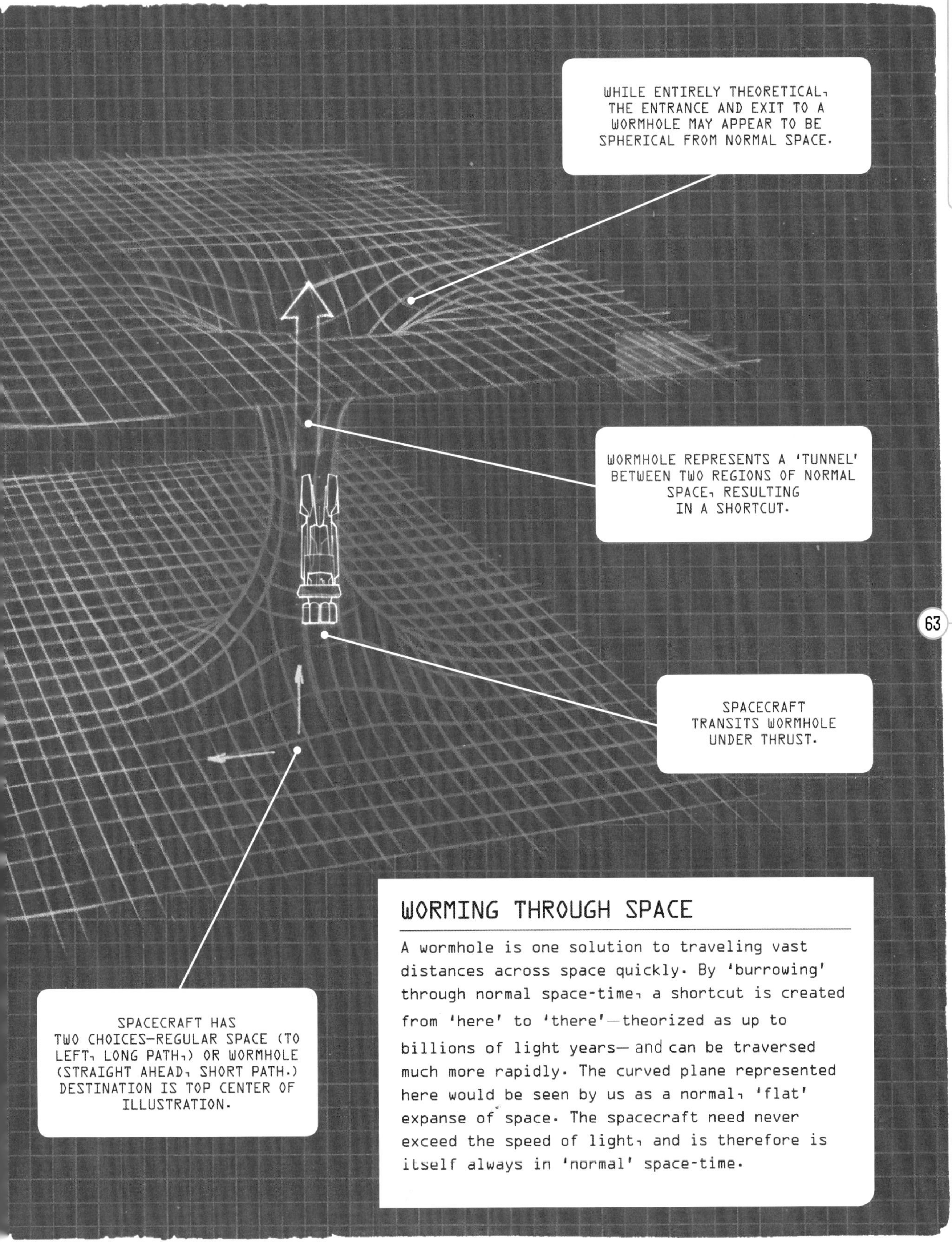

WORMING THROUGH SPACE

A wormhole is one solution to traveling vast distances across space quickly. By 'burrowing' through normal space-time, a shortcut is created from 'here' to 'there'—theorized as up to billions of light years— and can be traversed much more rapidly. The curved plane represented here would be seen by us as a normal, 'flat' expanse of space. The spacecraft need never exceed the speed of light, and is therefore is itself always in 'normal' space-time.

mathematically does not mean it can actually be accomplished, and physicists are not convinced that the math works. More seriously, others point out that this model also requires a negative energy to function. In addition, it has been suggested that, even if the system can be made to work, all the particles that the bubble would build-up in transit—the stuff it ran into while traveling—would be released in a burst as it reached its conclusion, thereby destroying that destination. But the truth is that nobody can be completely sure until even more complex equations are crunched, or we simply try it.

There have been attempts by dedicated teams at places like NASA's Johnson Space Center to research some of the elements of these ideas on a tiny scale in the lab. Sometimes equally small results are quietly leaked out, but they are usually followed by a storm of other scientists pointing out all the reasons it's probably wrong. There's a good chance that none of the research projects will work in the foreseeable future, but many 'experts' thought that airplanes could never fly and rockets would not work in outer space. So we will just have to wait and see.

One other new technology bears mention. It will not fly faster-than-light, but can move significantly toward *c*. This technique involves a spacecraft resembling a solar sail. These are vast, thin sheets deployed in space, usually made of something like Mylar. When properly oriented, they catch sunlight, and the propulsive force of all those photons smacking into the sail cause it to accelerate. Solar sails have been experimented with in Earth orbit and seem to work, but they are slow. They do, however, accelerate continuously and can build up a lot of speed over time. But what if you could ramp-up the 'wind' (in this case, solar wind) that powers them? Researchers have theorized various ways to beam

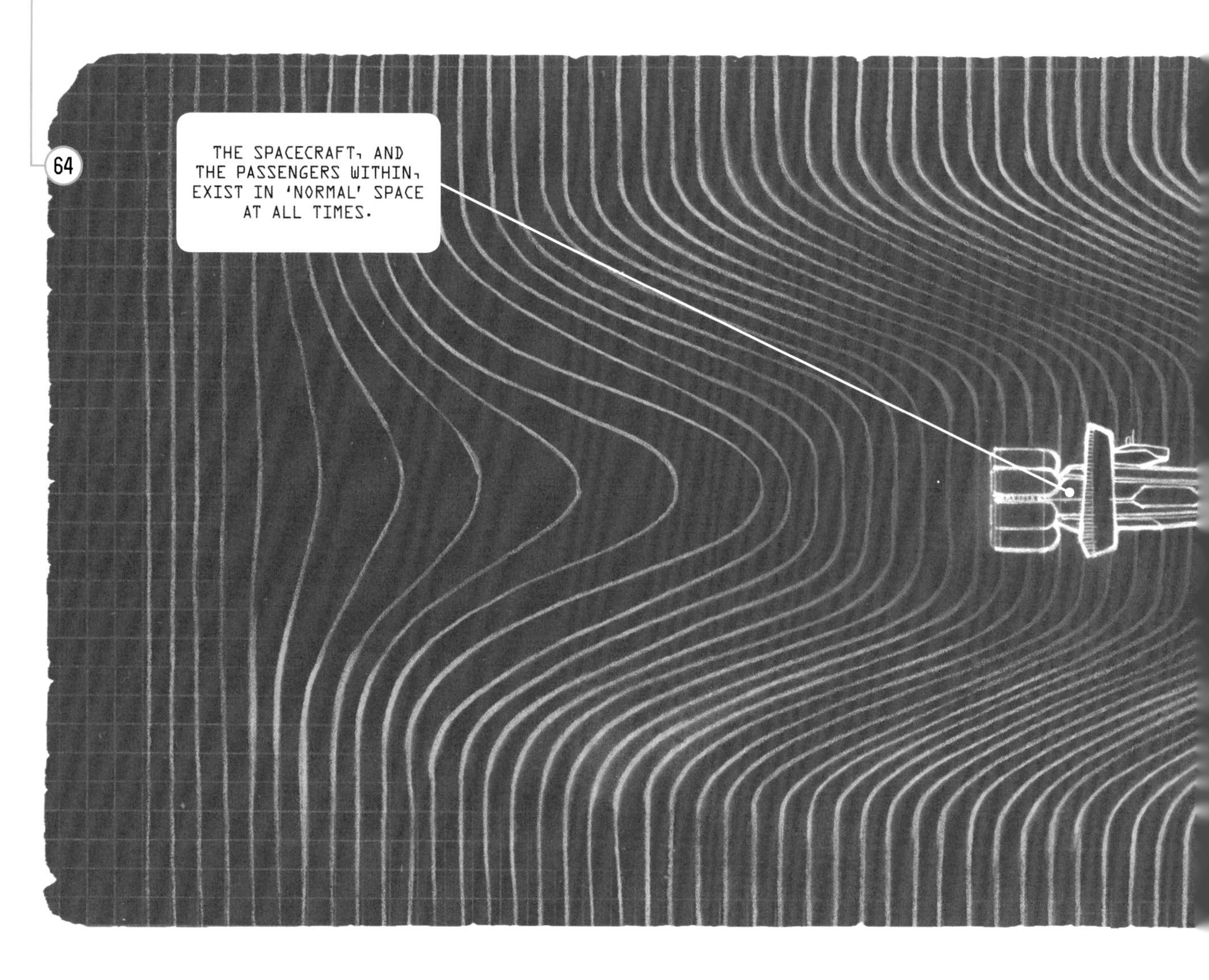

energy from Earth, or Earth orbit, to power a sail in space. This means that we are able to leave all that heavy fuel behind and only fly the cargo. A group of researchers at the University of California at Santa Barbara think they have most of the problems licked—at least in small scale. By shining a powerful laser into the sail, they have shown it is possible to get it to 'fly' merely by catching the photons from the beam. This kind of spacecraft, carrying something the size of the space shuttle, could reach Mars in just a month with a sufficiently powerful laser. In fact, they claim that their design would accelerate a tiny chip-sized craft to $^1/_3$ *c* in about ten minutes, using an amount of energy equivalent to that used to launch a Saturn V moon rocket. This electromagnetically driven sail may hold the key to crossing the vast distances in space in record time—such a sail, carrying a small robotic craft, could reach our old friend Alpha Centauri in 'only' seventeen years or so. But that's way better than a century with a toxic Orion nuclear-pulse drive. There's still a lot of work to be done on this concept.

There are, of course, a number of anti-matter designs on the drawing table, but until someone can figure out how to get sufficient quantities of the material, store it, and operate a drive with it, such notions remain in the realm of science fiction. But watch this space...for the incredible power of anti-matter may ultimately provide the best solution to this vexing problem.

Faster-than-light travel, in one form or another, remains the most promising way of truly reaching the stars. The physicists and engineers will keep chipping away at the edges of this problem—the physicists by looking for ways around Einstein's roadblocks to surpassing/bypassing the speed of light, and the engineers by seeking better, more efficient ways to fly at some reasonable percentage of it.

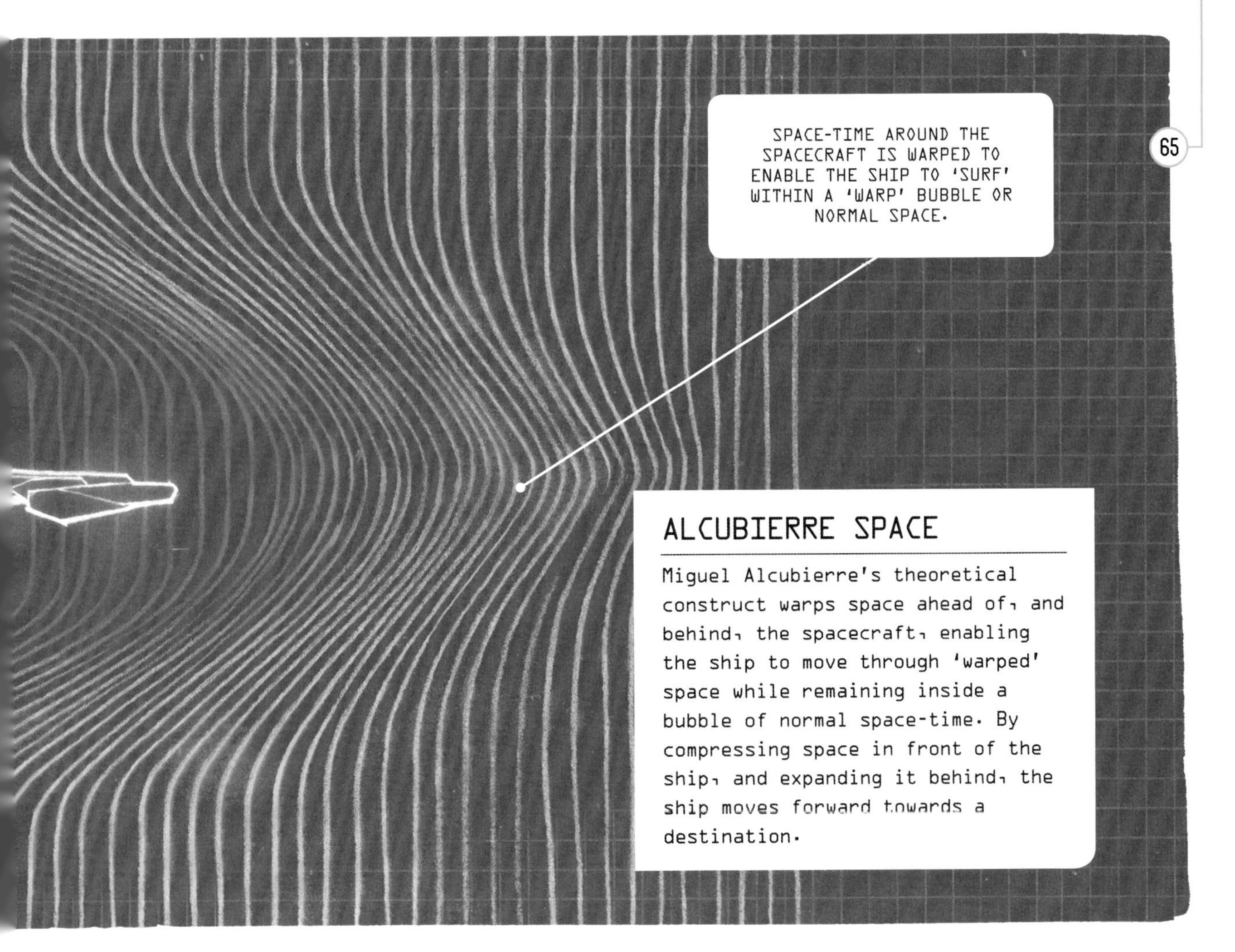

ALCUBIERRE SPACE

Miguel Alcubierre's theoretical construct warps space ahead of, and behind, the spacecraft, enabling the ship to move through 'warped' space while remaining inside a bubble of normal space-time. By compressing space in front of the ship, and expanding it behind, the ship moves forward towards a destination.

SPACE-AGED WAGON TRAINS: TRAVELING BY STARSHIP

Nothing has cemented the starship in our minds like the never-ending mission of *Star Trek*, created by Gene Roddenberry and now entering its sixth decade. Belonging to a genre that has become known as 'space opera', *Star Trek* was inspired by the Horatio Hornblower nautical adventure novels, in which earth-bound ships and their crews faced a strangely similar set of challenges.

TO THE STARS

The fact is, though, that when humanity progresses to using starships to travel to the stars, there is little chance they will look like the whimsical designs of *Star Trek* or *Star Wars*. Nor will they swoop and maneuver in the vacuum of space like Spitfire or Mustang fighter planes. It is unlikely that they will have artificial gravity throughout the ship. Unless there are some really big surprises out there, they are unlikely to be armed (though we should never discount our propensity for fighting among *ourselves* in space).

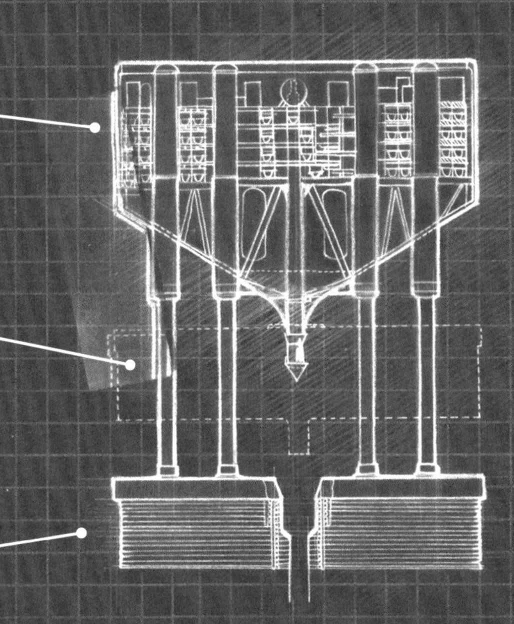

FUEL MAGAZINES HOLD HUNDREDS OF ATOMIC BOMBS TO PROVIDE PROPULSIVE FORCE. THE MAGAZINE AND PLATES AT BOTTOM AND TOP PROVIDE RADIATION PROTECTION TO THE CREW MODULE (NOT SEEN HERE, BUT WOULD BE AT TOP).

SECONDARY SHOCK ABSORBERS ATTACH THE PUSHER PLATE TO THE MAIN SPACECRAFT. AT CENTER IS THE EJECTOR TO SEND ATOM BOMBS THROUGH A HOLE IN THE PUSHER PLATE.

ENORMOUS PRIMARY SHOCK ABSORBERS COMPENSATE FOR THE INCREDIBLE RECOIL FROM THE CONTINUOUS ATOMIC EXPLOSIONS THAT DRIVE THE SPACECRAFT. BOMBS WOULD BE DETONATED AT INTERVALS OF A FEW SECONDS TO MINUTES, DEPENDING ON THE PHASE OF THE FLIGHT.

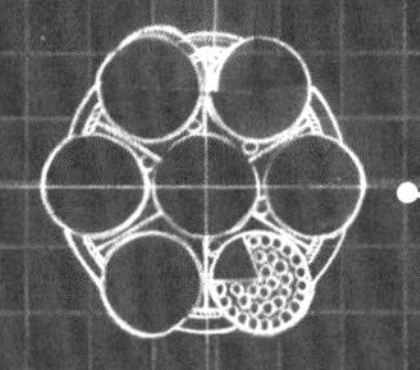

TOP VIEW OF PROPELLANT MAGAZINES, IN WHICH HUNDREDS OF SMALL ATOM BOMBS WOULD BE STORED. FROM HERE THEY WOULD BE DIRECTED THROUGH THE PUSHER PLATE (NOT SEEN) TO BE DETONATED A SMALL DISTANCE BELOW, PROVIDING MOTIVE FORCE.

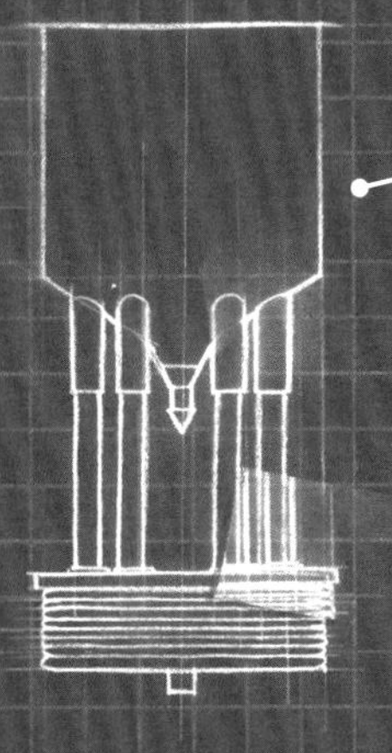

A LATER, SMALLER (33-FOOT) VERSION WAS PLANNED THAT COULD BE LAUNCHED ON A SATURN V ROCKET TO AVOID THE RADIOACTIVE FALLOUT OF THE ORIGINAL DESIGN.

GOING NUCLEAR

Project Orion would have constructed a giant spaceship, capable of carrying up to 150 people into space. It would have been much faster and more powerful than anything before or since...and would have needed hundreds of atomic bomb explosions to be launched. The study was ultimately shelved, though there is new interest in a space-only version of the spacecraft.

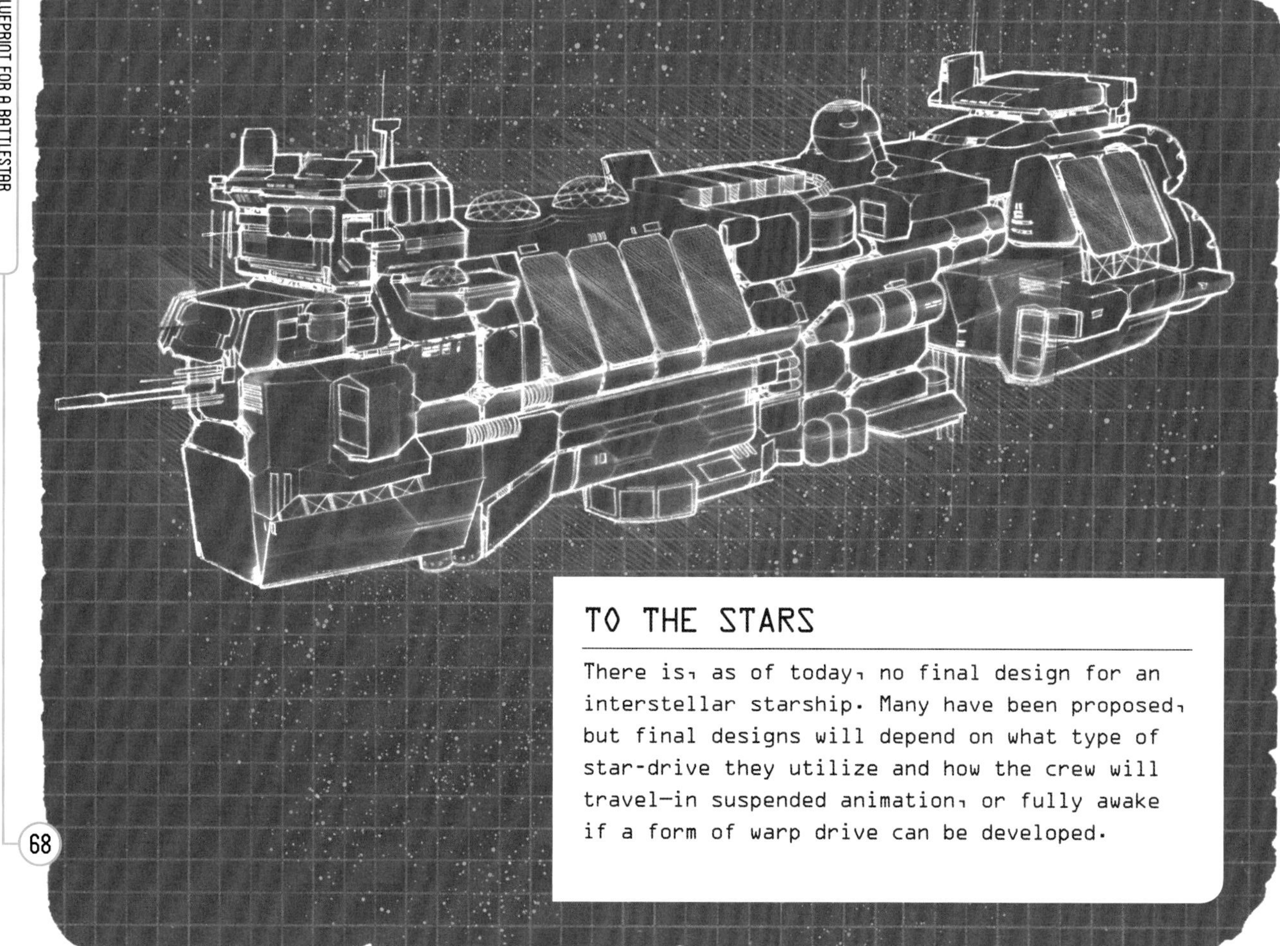

TO THE STARS

There is, as of today, no final design for an interstellar starship. Many have been proposed, but final designs will depend on what type of star-drive they utilize and how the crew will travel—in suspended animation, or fully awake if a form of warp drive can be developed.

In films, starships present a hugely glamorous image. But spaceflight, even with the invention of some kind of warp drive, is more likely to look like a cross between cramped economy-class airline travel and a military vessel. There will be few frills, interior space will be more about function than form, and there will likely be no gravity except in areas where it is created for health-maintaining purposes (and that is likely to be via centrifuge, not a magic-gravity-field). If you want to see what mass spaceflight will look like in its first iteration, a peek at the inside of the International Space Station is probably a good indicator.

But first, we have to have a reason to go into interstellar space, and the means to create vessels capable of doing so. To date, our major thrust into space with humans has been the Apollo program of the 1960s, and that 'spaceship' was a car-sized capsule with a small, paper-thin aluminium moon lander attached. There was just enough room for the three crewmembers to maintain their sanity for 8–12 days of high adventure. The space shuttle improved on this, with a separate flight and crew deck and about 2,500 cubic feet of space for the astronauts. The International Space Station gets us much closer to the probable reality, at about 13,700 cubic feet of habitable volume within an overall pressurized total of 32,000 cubic feet. Now all it needs is an interstellar star drive!

The first reasonably sound design for a large, high-speed spaceship (with interstellar potential) was the largest version of Project Orion, the magnificent—and slightly crazy—craft powered by atomic detonations, which we examined in Chapter 8. The 1950s design involved sliding multitudes of hydrogen bombs (for the racy interstellar version) from a central magazine, which were detonated a few hundred feet behind the protective, shock-absorbing pusher-plate. These detonations, at intervals of many seconds to many minutes, could build up high velocities in a hurry. And the really crazy thing? It probably would have worked, if it weren't for that one big design flaw: it would take 800 nuclear explosions, and the consequent fallout, just to get the thing into orbit.

With a mass of ten million tons for the 'Super Orion' (by comparison, the International Space Station weighs 450 tons) and a reasonable supply of explosives, a journey to Pluto and back was estimated (probably quite optimistically) to take just a year. Considering that we just spent a decade sending the unmanned New Horizons probe one-way to that icy dwarf planet, that's quite an improvement.

Orion eventually went nowhere, despite the millions of dollars ARPA (DARPA's predecessor) spent with the company that first developed the idea, General Atomics. NASA did some work on the concept. But between the engineering challenges and the political hot-potato nature of the nuclear propulsion system, the project was shelved in the early 1970s. Interestingly, designs similar to Project Orion (though on a far smaller scale), using modern materials with vastly improved efficiencies, are quietly being circulated today. If some of the mass of the monster spacecraft can be found already in space (metals from asteroids, for example), and an acceptably safe way of transporting nuclear materials to orbit can be arranged, who knows? And with the use of hydrogen bombs as a power supply, even the perceived evils of residual deep-space radiation would be minimized, given the amount of thrust produced via the larger output of the hydrogen explosion. There might eventually be an Orion-derived starship in our future.

There have been plenty of other designs for interstellar craft drawn up over the years. Most of them resemble Orion in overall form—crew up front, consumables and fuel storage in the middle, engines in the rear—but with different drive systems. We have already seen Boussard's stellar ramjet and Michael Alcubierre's 'warp' or space-time drive. There have been many other design entries into the superluminal (another term for faster-than-light) sweepstakes, but before we move deeper into the design of the ships, we should consider their primary cargo: human beings.

Humans make bad space travelers. We are basically five-foot-something bags of wet meat: all saltwater, flesh, and chemicals (the adult body is 60 percent water by weight.) These are things that don't mix well with the hard vacuum and extreme temperatures of space—when exposed to an open hatch, we swell up, the soft parts rupture or pop out, while the fluids boil and rush out of the nearest aperture, turning into useless ice crystals.

Humans are also somewhat parasitical, forming an endlessly hungry relationship with our environment. We breathe the air that Earth is kind enough to provide (but it has to be just the right mix of gases!), drink the water (but it must be fresh and clean!), eat the plants and animals, and leave our droppings hither-and-yon. Space is, in fact, hostile to all of these needs (save maybe the last one). So in the day-to-day business of staying alive, we humans are demanding and inconvenient beings to tote around the cosmos.

A few more factors count against us as spacefarers. We are essentially pack animals and we need other humans. Our psychology is delicate, and we don't do well alone for long stretches of time. That said, we're not too keen on being cooped up in small groups for too long either. Nor do we do well in radiation-rich environments, having been protected from space-borne radiation by Earth's magnetic field and atmospher since our amphibian days.

Then there's weightlessness to consider. Our bodies don't like it. After many millions of years walking erect and opposing the force of gravity on Earth, our bodies get very upset if they are deprived of its rigors for too long. In zero-gravity environments like space, bodily fluids, used to being pooled in the lower extremities, start to redistribute themselves throughout your body, making parts of you swell up that shouldn't. You experience a semi-permanent head-cold. Then, since your bones aren't working to hold you up against Earth's gravity, they go on a calcium-shedding holiday, passing it out through the urine and causing your skeletal system to become weakened—that is why the astronauts and cosmonauts you see returning from their multi-month cosmic adventures on the space station are rushed to stretchers and recliners. It's like reaching premature old age. Among the other odd things that occur in zero-gravity is that your eyes change shape. They elongate, throwing your vision out of focus and becoming less useful over time. And the changes, as far as the researchers can tell, are permanent.

With all these physical and psychological dangers to cope with, a key consideration in designing a spaceship to take us places is its ability to protect us from the very places we want to go, at least while in transit. Science fiction supplied ready answers for these issues over the years, and there are a set of pretty well accepted choices by now, which makes reading or watching space sci-fi so much more enjoyable (no struggles with a lousy oxygen mix or the trials of defecating in zero-g). The usual solutions range from suspended animation or hibernation, warp drives, protective force fields, and artificial gravity.

Unfortunately, like much of the science in science fiction, these things are harder to do in reality. Fortunately, we are making some progress toward dealing with these very real problems. Artificial gravity can be produced with on-board centrifuges; research in the International Space Station is teaching us how to minimize the untoward effects of zero-g; and various materials are being tested as workable radiation shielding. Metabolic slowing has been shown to be effective—and relatively safe—with mammals, and for the first time in decades serious money is being invested into alternative propulsion systems and even early experiments in warp-type transportation technologies.

One quick solution is to leave the fragile humans behind. That's what the Jet Propulsion Laboratory does as a Caltech-operated NASA field center, sending its robust robots to far-flung destinations, ranging from Mercury to beyond the solar system. Those machines are still affected by the hostile

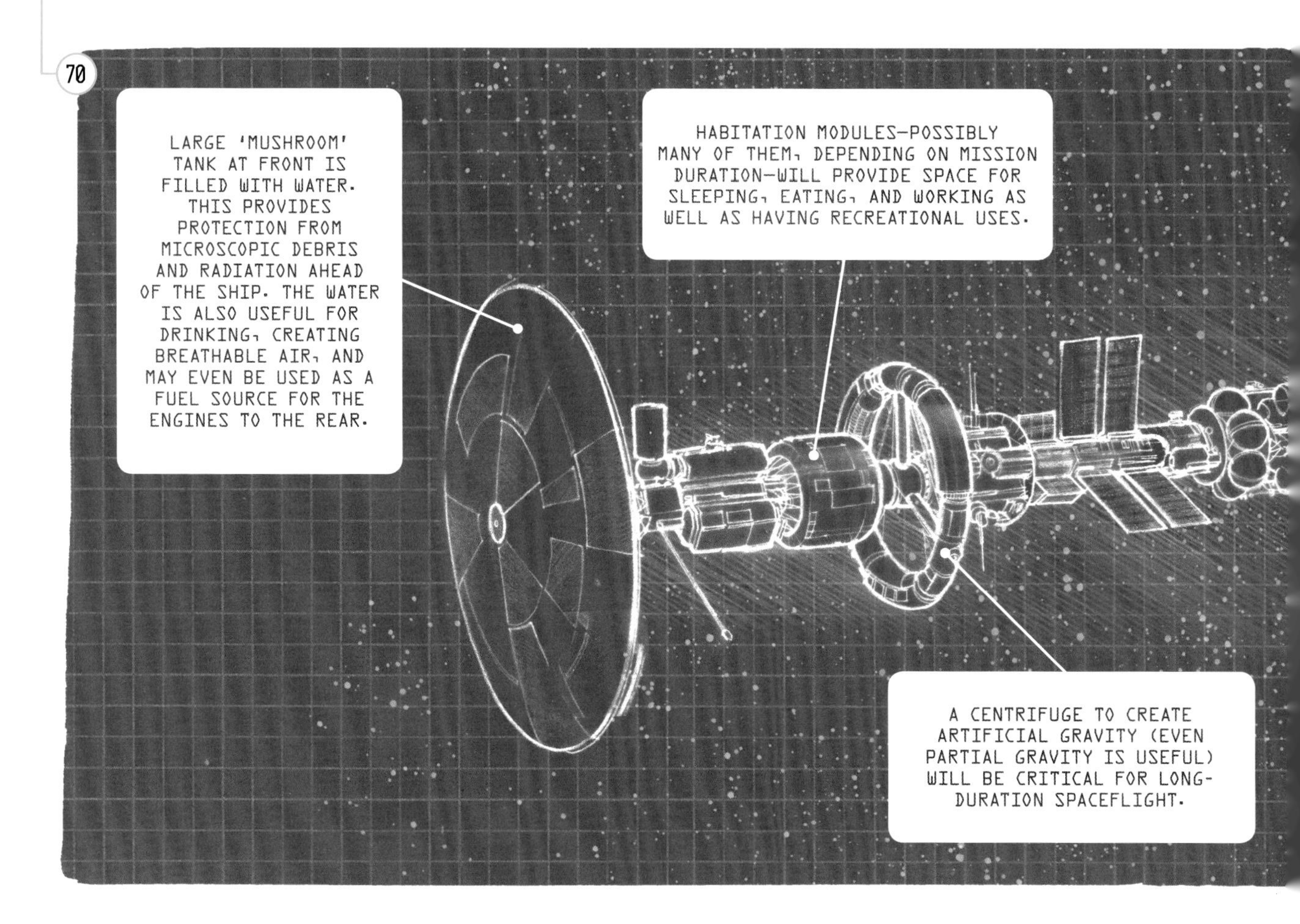

environments they work in; Martian dust eventually ruins electrical motors on the rovers, and over time, computers exposed to the toxic radiation out near Jupiter and Saturn will fail. But these robotic explorers have, in general, far outlasted their warranties, working selflessly into interplanetary overtime and asking for nothing but radio instructions in return.

But this is an intrinsically unsatisfying solution. People want to *go* places, and space is no exception. While the scientific returns from robotic probes are vast and gratifying, it is just not the same as sending people there. That firsthand, this-is-what-it's-like-to-stand-on-planet-X narrative, thrills the masses. We crave adventure and exploration, even if it is secondhand for the vast majority. That's what makes us an exploring species.

One answer comes from the newish discipline of transhumanism. It's a vast field, of which freeing the brain from the human body is just one facet. As we will see in Chapter 16, creating a cyborg around a human brain (which is, after all, still the fastest and most capable computer known) sidesteps the vast majority of the problems associated with long-term spaceflight. There would still be issues of radiation exposure, but a brain is much more easily shielded than an entire body. There is still the question of life support, and the brain is indeed a hungry thing, requiring about 30 percent of the calories we consume to do its job. So you have to feed it well, but it gives back a lot in return.

A sand-grain piece of brain tissue contains 100,000 neurons with a billion synapses all chatting with one another, so it's a really effective computer. It just needs a better, more robust support system. But this kind of machine/brain symbiosis has not yet been achieved, and so we don't have little human brains sitting atop space robots just yet. There are also a plethora of ethical issues around creating human/machine hybrids, so if there is work proceeding in that direction, the researchers are keeping it quiet.

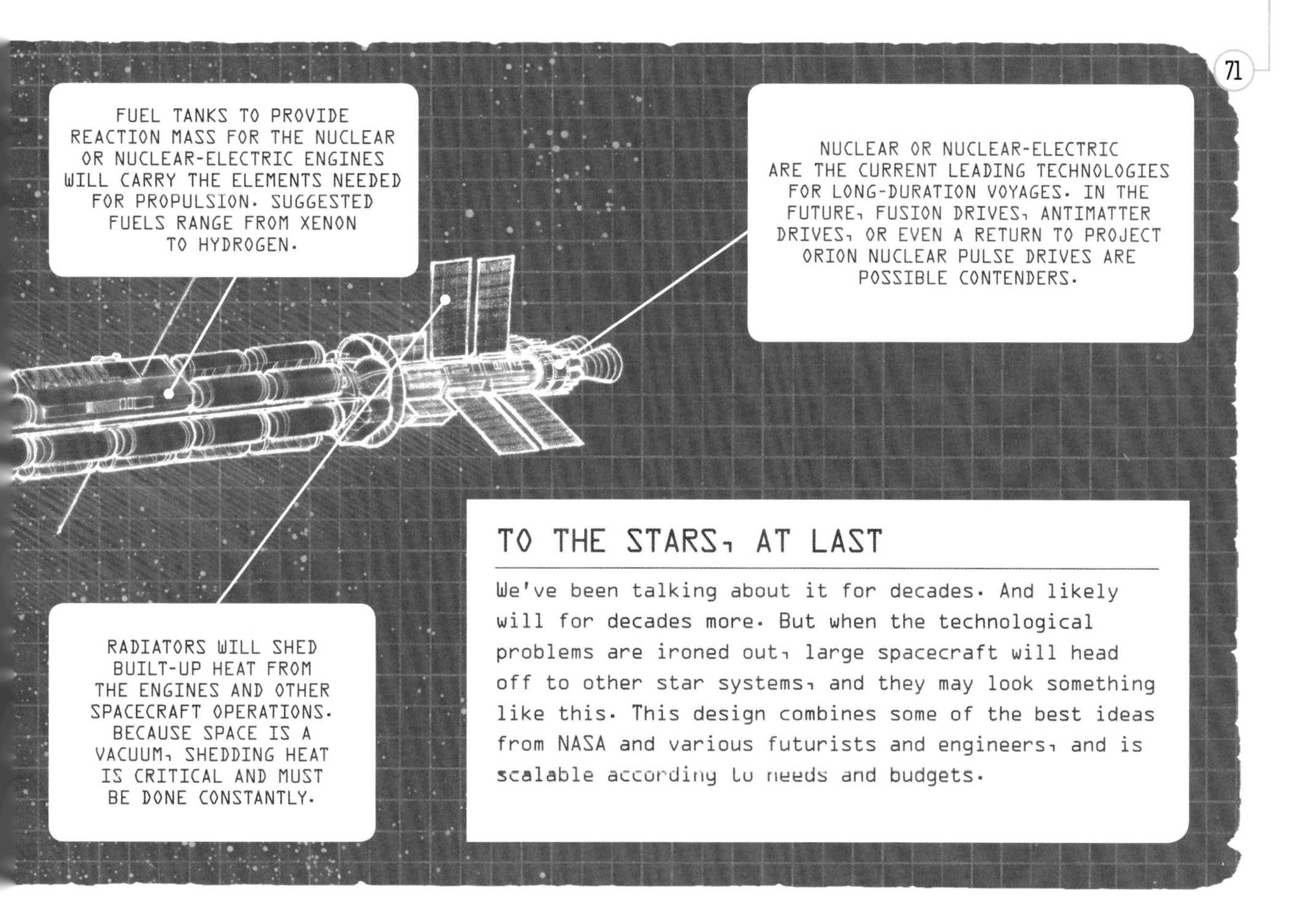

TO THE STARS, AT LAST

We've been talking about it for decades. And likely will for decades more. But when the technological problems are ironed out, large spacecraft will head off to other star systems, and they may look something like this. This design combines some of the best ideas from NASA and various futurists and engineers, and is scalable according to needs and budgets.

Human augmentation is another possible way to enable long-distance space travel. Parts of the body could be bypassed and overridden to minimize the damages incurred by spaceflight, but this also is tricky business. Evolution has done a pretty good job of setting up how the body works, and when you mess with that, it's a bit like unraveling a ball of yarn—easy to do, but hard to set right. A 1960s scientific study pondered 'doping' astronauts to make them better at spaceflight, but there are probably other ways of 'hardening' the body to better survive the experience. Even so, anything that actually works is a probably some way off.

How about hibernation? This is an intriguing area of research, and is, after many decades of experimentation, showing some promise. Gone are the days of freezing heads to preserve us for revival at some future date; cells are just too fragile, and the water within them (there's a lot of it) does not like being confined when it freezes. The resulting ice crystals do awful things to the tissue they are trapped in.

Nature often provides the best answers for how to accomplish things and so scientists have been studying animals that can hibernate for clues. From small organisms like frogs that can survive freezing and thaw back to life, to hibernating bears who sleep over the winter, then emerge ready to hunt, there are natural mechanisms that slow bodily functions—a lot. The trick will be to apply this to people. Research is currently being conducted on a species of lemur, the only primate known to hibernate, and how their ability to slow their metabolism, lowering body temperature, and blood pressure, might be applicable to humans.

As popularized in science fiction, future starships may place most or all of their crews into deep sleep, keeping the human cargo in fine fettle via lowered metabolic rates with base-level nutrition and life support. Let the artificial intelligence (see Chapter 5) run the ship while the humans sleep it off, then allow them—with all their nagging physical and emotional needs—to emerge unscathed at the end of the trek.

Another approach to the question of human survival in space has been to preserve the human crew via generational starships. While this is less complicated with regard to bioscience, it is far more so in terms of life support and spacecraft design. Instead of a relatively sleek, massive refrigerator warping through space with slumbering humans aboard, we would have a vast space-faring city and all that is associated with keeping generations of people alive and both physically and psychologically healthy. This approach would also need a sufficient mass of human breeding stock to allow subsequent generations to produce viable offspring—estimates of the minimum number have ranged from dozens to hundreds (though this may be reduced via future developments in genetic engineering or stored genetic material).

Generational starships would also require a well-stocked hospital with either human or robotic doctors on standby to keep the crew in top form. Gymnasiums—complete with some kind of artificial gravity—would be necessary for the physical health of the crew, along with 'green zones' (artificial parks with engineered bits of nature) to optimize their mental state. On-board farms would be necessary to provide food for the crew, unless all the nutritional requirements are synthesized, which to date has not provided very satisfactory (or tasty) results. Then there are sleeping quarters (complete with artificial day/night cycles to preserve natural circadian rhythms), recreational areas, science labs, shops for maintenance and repair—the list goes on. These ships would also need a form of government, which would not necessarily be a democracy, and a legal system with some kind of police force to uphold it. The practical problems are huge.

Like most starship concepts, generational starships would have to be constructed in space, largely from raw materials found there on asteroids and moons. It would be a massive technological undertaking. And the closest we have reached to simulating the needed self-sustaining environmental technology on a large scale is the Biosphere II project in the Arizona desert, which ultimately failed. The design of the Biosphere complex introduced a shortage of oxygen and there were complications balancing the internal ecosystem, but these shortcomings were ultimately understood and the challenges are not seen as insurmountable. It just takes time, dedication, and lots of money.

Of course, both these types of ships—hibernation or generational—remove the crews from contact with their place of origin, the Earth, and the society that created them. The voyage to our nearest stellar neighbor, Alpha Centauri, will take over four years even at the speed of light, and much longer with known technologies (as discussed, even a ship like that central to Project Orion would take well over a generation to arrive; perhaps hundreds of years). Messages would take an increasingly long time to make the one-way trip to or from the spacecraft, making two-way communication impossibly unwieldy.

The long travel times could cause a severe case of 'arrival shock,' as dramatized in the story *Far Centaurus*, written by A.E. van Vogt in 1944. In this tale, a group of settlers set out for the nearby (in relative terms) star system Alpha Centauri, only to discover upon their arrival that subsequent technological development has surpassed them. Far faster starships have arrived ahead of them, even though they had departed decades later, and have colonized the system. The pioneers in the first interstellar craft had slumbered while the others passed them by. They were truly people 'out of time,' who experienced great difficulty fitting in to the new society that had left them in its wake. As pointed out by futurist Robert Forward, the crew might end up having to go through interstellar customs upon their arrival at the distant star system. It is an interesting scenario.

Our current knowledge does not indicate that there are inhabitable worlds at Alpha Centauri. The nearest possible candidate might be in the planetary system orbiting a star called Tau Ceti, which is more than 11 light years away. Two planets, Tau Ceti e and Tau Ceti f, are so-called 'super-Earths' weighing in at a few times Earth's mass and, we think, within the 'habitable zone,' a distance from the host star that might make the planet suitable for life of some kind. Whether or not that means suitable for humans is uncertain. We still need to find a source of water and, ideally, a breathable atmosphere. Tau Ceti is just one possible candidate and there are innumerable others, but the more we learn about exoplanets (planets in other solar systems,) the fewer of them seem to look habitable. The Kepler Space Telescope has performed admirably since 2009, scouting out exoplanets for further investigation. The investigation of these bodies is in very early stages; it is simply too early to say with any confidence what will turn up.

Of course, if light-speed or warp drive do become a reality in the foreseeable future, it would be a game-changer. A starship capable or traveling at the speed of light or *c*, would take something over 12 years to reach Tau Ceti, discounting the time taken to accelerate to *c*, and to decelerate to the target world. It would be far less for the crew, due to relativistic time dilation.

Because of the time dilation, the crew would need only a fraction of the 12+ years-worth of supplies. But the large masses of fuel the ship would need to carry might be a problem. This would be less so, if the ship used an anti-matter drive or something like Boussard's interstellar ramjet (see also Chapter 8). And not at all if the vessel used beam propulsion, in which the spaceship is a target for a hugely powerful energy source—a laser or microwave beam aimed from our solar system—when the fuel is effectively left at the source. Our starship goes from being a massive fuel-tank-with-passengers to a far smaller craft, akin to a huge railroad car which is carried along by an external propulsion source.

And what about the warp-drive ship? NASA and DARPA have invested some funds in the subject, and one result was a starship design named—unsurprisingly—*Enterprise*, which is about the size of a small naval vessel. A lot of that mass is for the power units (although these do have yet to be invented.) Using Alcubierre's warp-drive principles (another type looked at in Chapter 8,) the craft could make the journey to Alpha Centauri in just over two weeks, and Tau Ceti within a couple of months. That's a lot better than Project Orion or even our just-under-*c* spacecraft— so, if only we can bring Alcubierre's ideas to reality, they probably present us with our best bet for reaching the stars.

THEY DON'T WORK ON WATER: REAL-LIFE HOVERBOARDS

When we saw the first hoverboard in 1989's *Back to the Future II*, it just looked so...*right*. In the second movie of the franchise, Michael J. Fox's character is transported to the then-unthinkable future (the year 2015!) Like flying in a dream, his hoverboard adventure seemed like something that simply *should* be.

A SKATEBOARD WITHOUT WHEELS

2015 came and went with no hoverboards in sight—at least, not real ones. There is a two-wheeled Chinese version with a patchy safety record, but real hoverboards are, like the jet pack, not quite yet on the shelves. There have been many attempts, and a few near misses, but to date, that's about it. The brilliant fictional creation of director Robert Zemeckis works only with a green-screen rig, wires, and a heavy application of visual effects.

The idea of a hoverboard as a skateboard without wheels, which uses technology to negate the force of gravity, appeared in at least one science fiction story before the *Back to the Future II* movie In a 1967 novel *The Hole in the Zero* by M.K. Joseph, we meet a character who 'chattered on about girls and riverside dance halls and hoverboard skating...'

But what about the real thing? We know *what* it is supposed to do, but *how*? The film never really explains how it's supposed to operate. We assume that it is some kind of magnetic-levitation device, but there's the little unexplained issue of how the physics works.

In the real world, two basic types of hoverboards have been demonstrated. The first is usable only over specially built surfaces, which give the hoverboard something to push against magnetically. The second type works on regular (but very flat) surfaces, but is really a miniature hovercraft, using

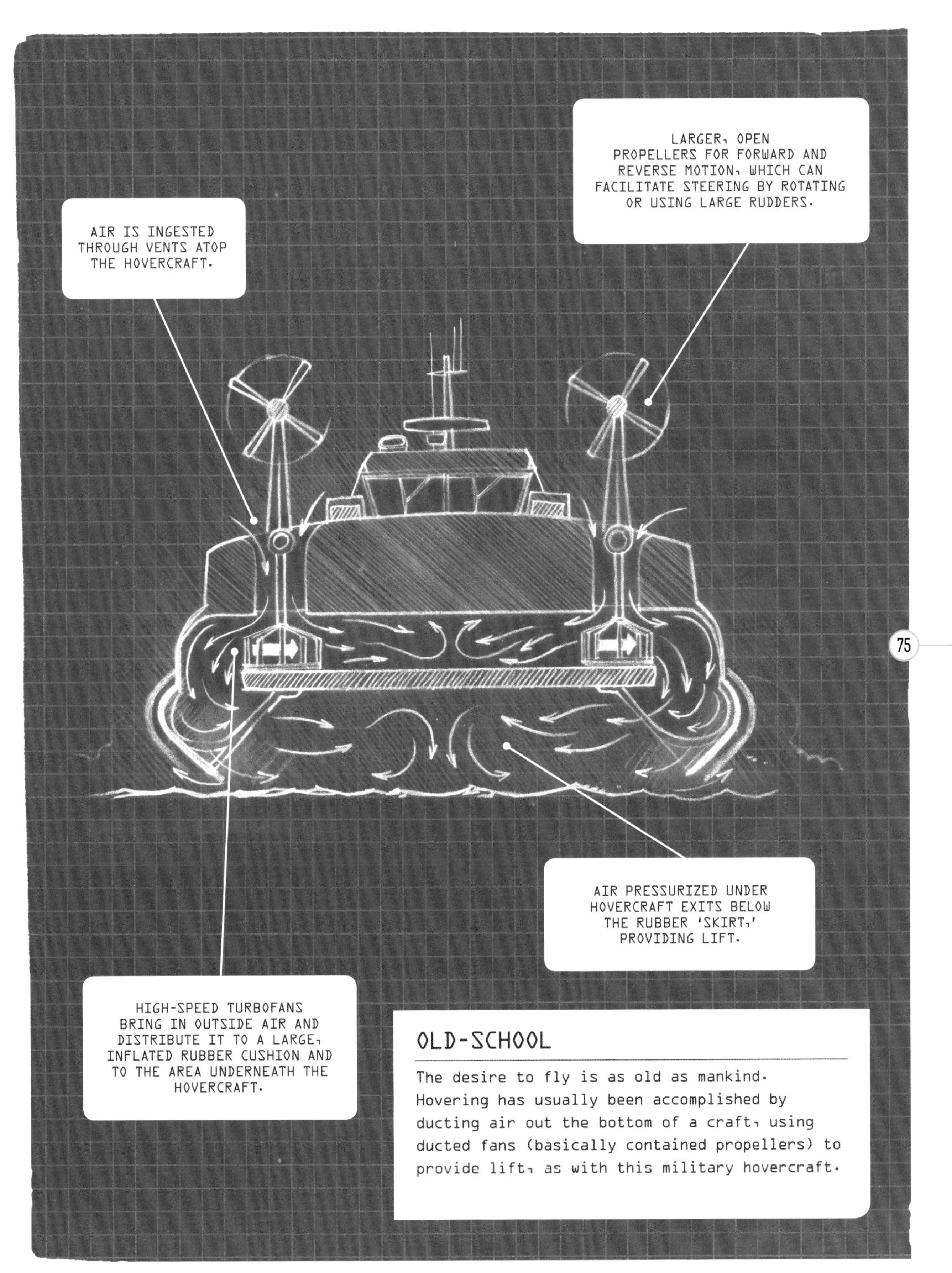

OLD-SCHOOL

The desire to fly is as old as mankind. Hovering has usually been accomplished by ducting air out the bottom of a craft, using ducted fans (basically contained propellers) to provide lift, as with this military hovercraft.

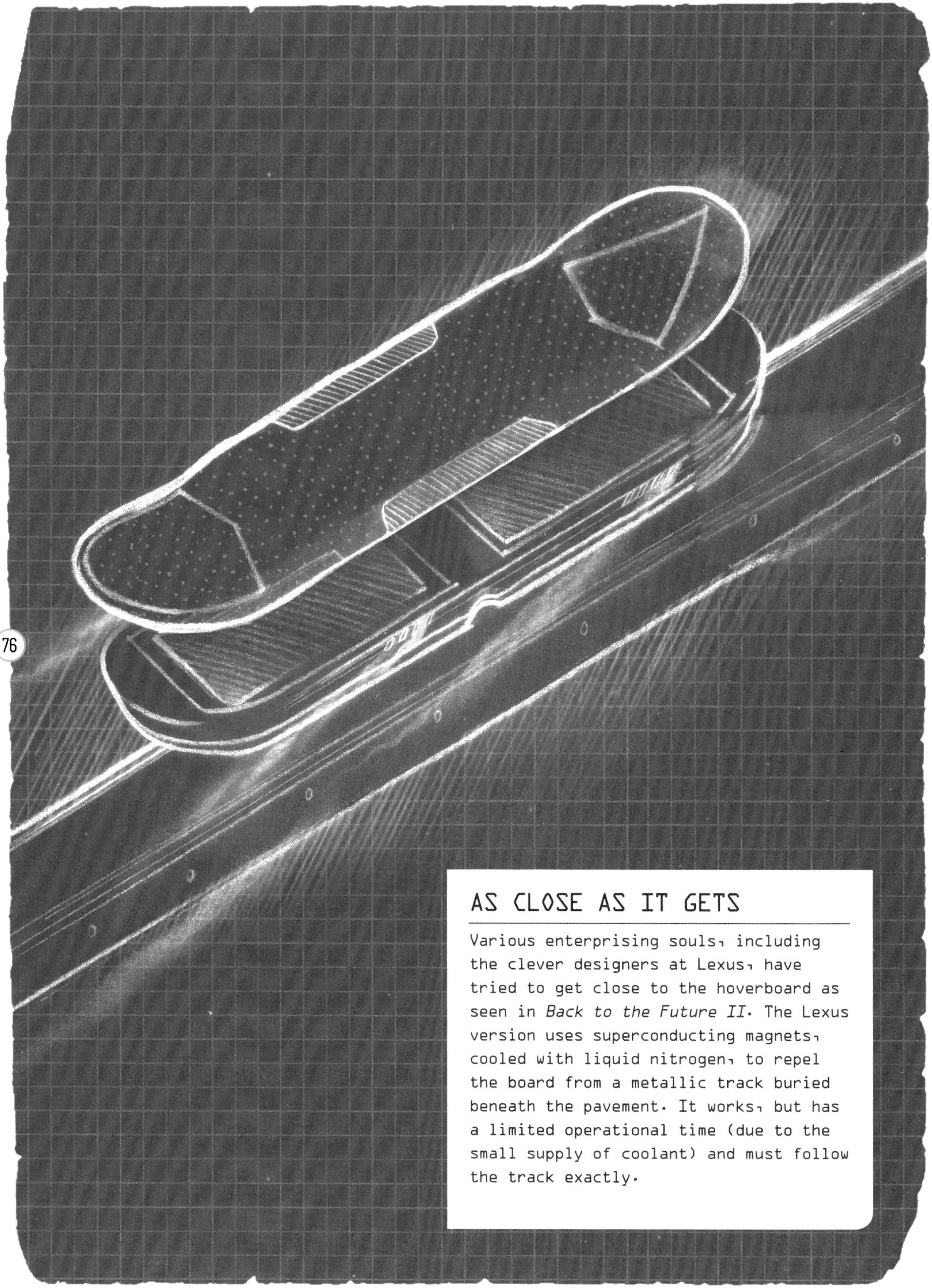

AS CLOSE AS IT GETS

Various enterprising souls, including the clever designers at Lexus, have tried to get close to the hoverboard as seen in *Back to the Future II*. The Lexus version uses superconducting magnets, cooled with liquid nitrogen, to repel the board from a metallic track buried beneath the pavement. It works, but has a limited operational time (due to the small supply of coolant) and must follow the track exactly.

fans to loft the deck on an air cushion (making them large and noisy).

One example of the fan-driven version is the Omni Hoverboard. It's not compact—the first version was about the size of a small door and used electric fans to fly, making it enormously noisy. While it is a bit ungainly looking, and does not fly for very long, it has one huge advantage over a real hoverboard: because it is propelled by fans, which create an air cushion below it, angling the Omni results in forward motion. It also does not care what it flies over; the machine will scoot over water just as happily as pavement or any other smooth surface. The prototype of a second-generation version managed to fly for almost 1,000 feet, a record for now.

A similar product called the ArcaBoard has been demonstrated in Europe. It looks like a long tombstone and uses 36 ducted fans to generate lift —the same principle as the Omni, and it works about as well. With a thin rider, it can fly for five or six minutes over anything flat. It weighs 180 lb in its current form, so carrying it around needs a vehicle.

In 2014, a company called Hendo announced a hoverboard that uses what the inventor calls 'Magnetic Field Architecture' to levitate. The device is packed with batteries and electromagnets, and looks (and behaves) far more like the hoverboard in the movies than the fan-driven ones. These magnets create a constantly shifting magnetic field when electrified, and this in turn induces a repulsive force in the the specially prepared surface below (which has to be copper clad and fairly smooth). While the copper itself is not magnetic, it is what is known as an inductor, meaning that when exposed to a magnetic field, it in turn creates one of its own. So when the Hendo hoverboard is powered-up over the copper surface, the magnetic field created 'pushes back,' creating a lifting force in the board. Aluminum and a few other metals with high electrical conductivity materials should also work.

But by far the most complex hoverboard comes from Lexus, the luxury car maker. In a hugely expensive publicity stunt, Lexus engineers developed a hoverboard that uses superconductors to accomplish levitation. These need to be really cold to function properly, so the Lexus unit trails a wispy stream of vapor behind it as it operates, courtesy of the tank of liquid nitrogen that chills the magnets within. This version needs magnetic tracks below it to work, so while the demo videos make it look as if it is freely flying over a skate-park like course, the rider is actually following magnetic rails underneath the surface.

The Lexus hoverboard looks a lot like the one from the *Back to the Future* movies and moves like a dream, at least for a few minutes until the liquid nitrogen supply runs out. But with its -320°F superchilled 'fuel' and the need to follow embedded rails, it is really more of a stunt than a practical design to build upon. Incidentally, it also takes a bit of training to ride well. In the two-minute video produced by Lexus, the resilient young people who try it fall off left and right until finally, a few daring souls get it right and ride to glory.

If real hoverboards haven't yet reached *Back to the Future* levels of technology, there was always the infamous hoverboard hoax that fooled many people. In 2014, a fictional company released a fictional video about a fictional hoverboard called the HUVr. Actor Christopher Lloyd (Professor Emmet Brown from the *Back to the Future* movies) and skateboarding legend Tony Hawk, were featured in the video. The five-minute show has Lloyd introducing the new invention, and says in a title card, 'The following demonstrations are completely real.' Then Hawk, with hands shaking and voice a-quiver, soon unpacks the HUVr, which looks just like the movie version. Within moments people are flying right and left, performing snappy turns and gnarly stunts galore. It's a beautiful video, and a skillfully portrayed lie. After the piece hit the Internet, a few sharp-eyed viewers noticed some visual inconsistencies, including the dynamics of how it was ridden, as well as some visual-effects problems, such as the incomplete digital removal of the suspending wires.

So will we ever see a real, practical, simple, non-liquid-nitrogen-fueled hoverboard zooming through a park near you? The simplest answer is: you may, when we have affordable, reliable room-temperature superconductive magnets. Such a contraption would still need magnetic tracks or specially prepared surfaces to operate, but it would be a blast to use. Otherwise, we would seem to be stuck with noisy fans (unless someone suddenly invents anti-gravity coatings).

ARE WE THERE YET?: THE FLYING CAR

Many of us first saw the flying car in its somewhat mutated (and misshapen) form in the animated cartoon show *The Jetsons*. Of course, that program also had a wheeled and fussy robotic maid, and life was run entirely via single pushbuttons—it was all part of the joke. But the idea stuck with us. Sometimes, it seems as if we're trying way too hard for something we don't really need. Yet, as self-driving, satellite-navigated conventional automobiles come close to fruition, maybe self-flying cars deserve a second look.

THE FLYING AUTO

The idea of a car that could fly predated *The Jetsons* by decades. The idea of a car that could fly is, in fact, as old as the car itself. A small group of enthusiastic (some might say 'fanatical') inventors were the founding fathers of the flying car, and have continued to invest much time (and expense) in their ongoing efforts to make them a reality. Most of these efforts were at best partially successful.

One of the earliest serious attempts was by an aircraft designer named Glenn Curtiss, whose

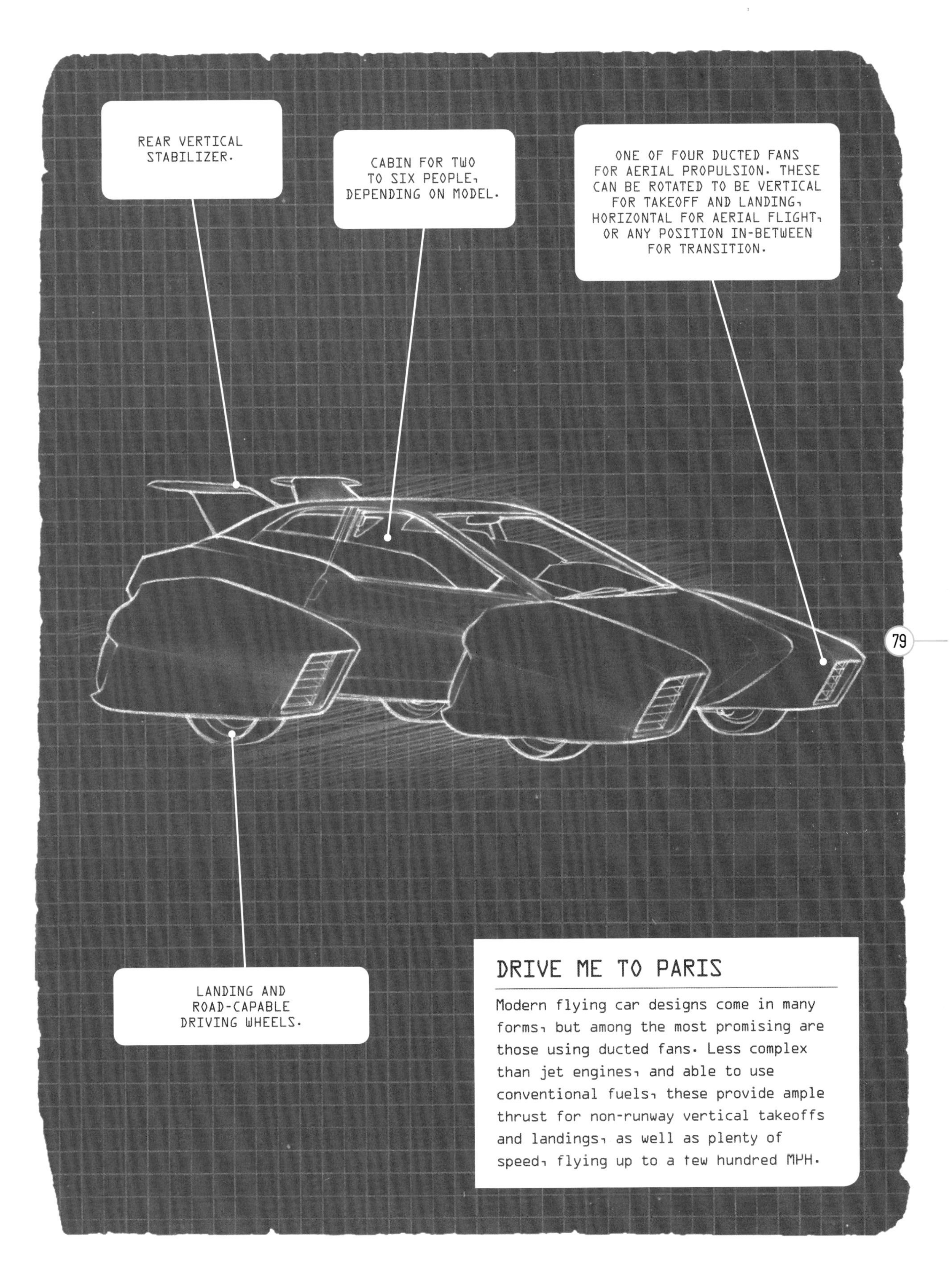

DRIVE ME TO PARIS

Modern flying car designs come in many forms, but among the most promising are those using ducted fans. Less complex than jet engines, and able to use conventional fuels, these provide ample thrust for non-runway vertical takeoffs and landings, as well as plenty of speed, flying up to a few hundred MPH.

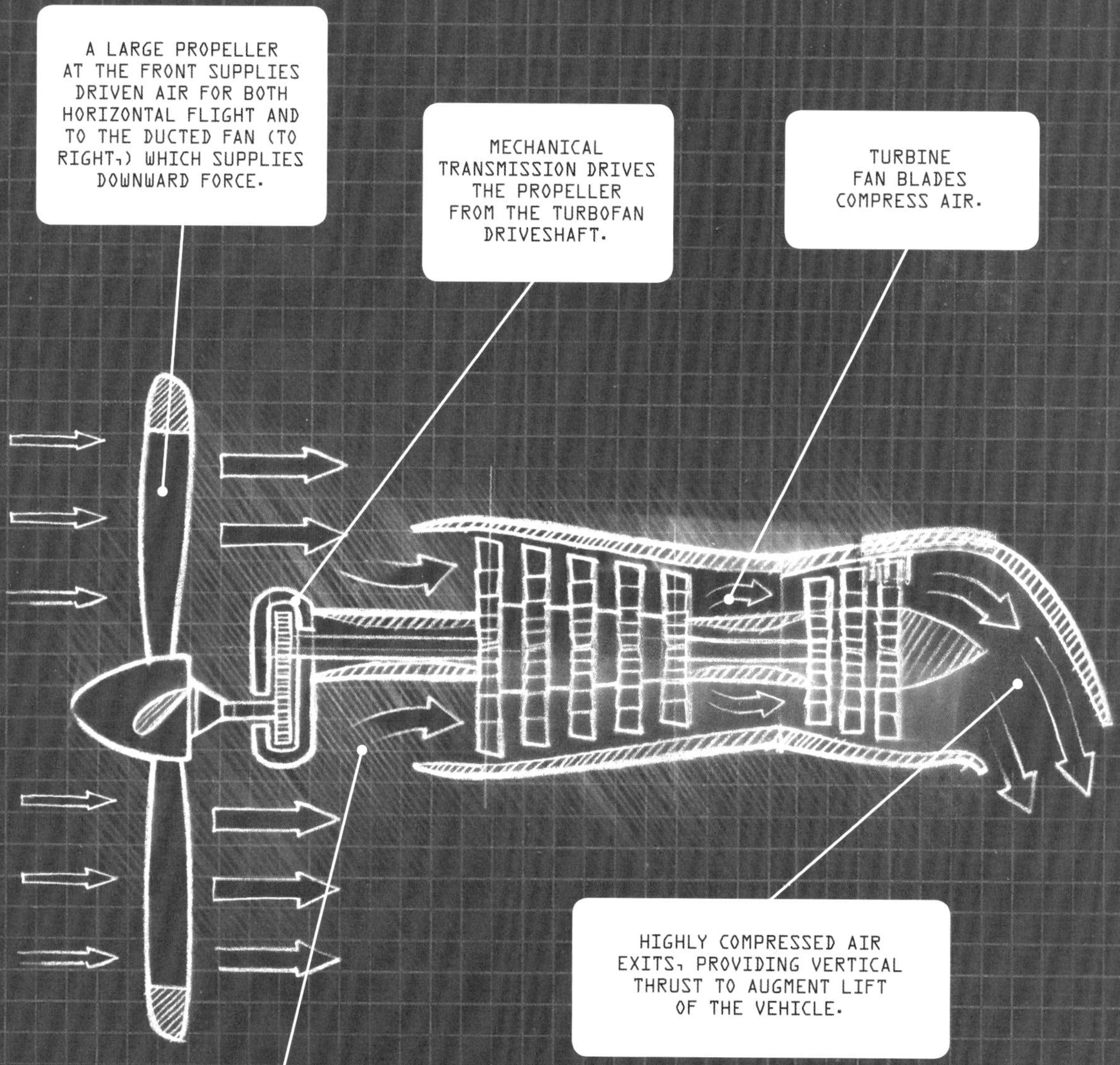

A LARGE PROPELLER AT THE FRONT SUPPLIES DRIVEN AIR FOR BOTH HORIZONTAL FLIGHT AND TO THE DUCTED FAN (TO RIGHT,) WHICH SUPPLIES DOWNWARD FORCE.

MECHANICAL TRANSMISSION DRIVES THE PROPELLER FROM THE TURBOFAN DRIVESHAFT.

TURBINE FAN BLADES COMPRESS AIR.

HIGHLY COMPRESSED AIR EXITS, PROVIDING VERTICAL THRUST TO AUGMENT LIFT OF THE VEHICLE.

EXTRA AIR, DRIVEN BY THE PROPELLER, ENTERS THE TURBOFAN HOUSING.

USING THE FORCE

For a flying car to work without a long runway for takeoffs and landings, a large mass of air must be moved through a compact assembly to provide thrust. This diagram shows one version of how this might be accomplished via a front propeller and turbofan compressors, with high-velocity air exiting at the right, ducted downwards to provide lift for the vehicle.

Curtiss-Wright Aircraft Company would later build the P-40 fighter of World War II fame. His 1917 Autoplane was perhaps more airplane than car, but it was an attempt to mix the two when the automobile was still in its infancy. It managed a successful test flight (more of a short hop, really,) and Curtiss moved on to more air-worthy designs. He was too early to the game...cars had a long way to go towards maturation, as did airplanes. Marrying the two technologies at this early stage was unlikely. The Autoplane did not catch on.

In 1940, stories appeared in the press about an American inventor named Jess Dixon, who created what he called the Flying Auto. It was really more of a frame with two stacked helicopter blades atop it powered by a whopping 40 hp engine. Not much else is known, but a few photos of an allegedly successful test flight remain. Unsurprisingly, it did not catch on.

In 1949, the Taylor Aerocar flew onto the scene. This was the most serious contender yet, with six prototypes built. The Aerocar had an enclosed body that looked a bit like a squashed egg with insectoid wings folded back when in road configuration. When it was time to fly, the wings were repositioned forward, attached to the car, and a propeller affixed to the back, allegedly within five minutes. The unwieldy contraption passed certification by the FAA (then called CAA,) and was capable of 100 mph in the air and 60 mph on the road. It was expensive, at a then-whopping $25,000, and never went into general production. It did not catch on.

In the 1970s, two companies attempted to build flying cars using already established production vehicles. The first was the MIZAR, a mashup of a Cessna Skymaster (a unique pusher-propeller plane used in Vietnam) and...a Ford Pinto. Unfortunately, the prototype crashed, resulting in the death of the inventor, and it, too, did not catch on.

At about the same time, a flying car appeared in the 1974 James Bond movie *The Man With the Golden Gun*. In the movie, the villain escapes by converting his 1974 American Motors Matador into an airplane. It was an unappealing hybrid visually and was only capable of take-off and a short flight of about 1,000 feet, and so it, too, failed to catch on.

The flying car went into well-deserved oblivion for a while, but the advent of the 21st century has brought the notion of mass-produced flying cars back into vogue. Composite materials, some lighter and stronger than steel, have been game-changers. Lighter and more powerful engines are available. GPS and computerized operation may actually make flying cars safe to use. And finally, traffic in modern cities has become so horribly congested that it might be worth the $200,000, and, up prices to soar above it all.

At least ten companies are striving to develop a road worthy flying car internationally. Those that do not need to comply with US crash testing will have an easier time of it, and prices may be correspondingly lower. Various schemes to achieve the needed lift have been employed. Some are in the end-stages of prototype development; others are likely vapor-gadgets, which will never reach production. But all are intriguing, and the quirky nature of some developers comes with the territory.

The Terrafugia Transition is built by a US company founded by a pack of MIT graduates, so they do have engineering credentials. They have a number of standing orders for the $300,000 vehicle, and while lagging behind their initial delivery promise of 2015 or so, are progressing nicely. The Transition looks like a small airplane with wheels, with a propeller to the rear and wings that fold out in a gull-wing configuration in less than 60 seconds. It flies like a traditional airplane, so will need takeoff and landing runs wherever it flies and lands. The designers claim that it will fit in a normal parking space or garage, and can fly for 450 miles on regular gas! They are currently working on a completely different concept, the TF-X, that uses electric motors to take off vertically and then transition to horizontal flight, carrying up to four passengers. The company has won research contracts from the FAA, so one day soon we may actually see their vehicles in operation.

A Dutch company has developed the PAL-V, which uses a somewhat different approach. It is akinto a hybrid of an enclosed motorcycle and gyroplane. A gyroplane looks itself like a mix between a car or cycle and a helicopter, except that the chopper blades are unpowered—forward motion from the rear-mounted propeller makes them spin. The PAL-V is a three-wheeler that comes in a number of configurations, but in general has a very short take-off and landing requirement, a ceiling of 4,000 feet and speed of 112 mph. The rotor blades fold completely back when in road configuration, tucking in to the rudders at the back. The PAL-V one has completed preliminary testing, and could be available soon—though there are compliance issues to be worked through. If you're going to have cars flying overhead, you want to make sure they stay in the air, right?

Certainly the most likely contender for the 'as we saw it in the 1930s' award is the Moller Skycar. Paul Moller, the inventor of the futuristic looking craft, says he's been working on flying car designs for more than 50 years. There have been a number of prototypes; the one seen at the start of this chapter is the M400. The first tethered (roped to the ground for safety) demonstration was in 2003. It is a VTOL (Vertical Take Off and Landing) design that uses ducted fans for lift—louvers pivot downward for launches and landings, then sideways for vertical flight. The capacity is listed at between one and six passengers, and the claimed maximum altitude is 36,000 feet with a speed of over 300 mph, but as the real thing has not yet materialized, all of this is pure speculation on the designer's part. Despite many tens of millions of dollars invested, and promises for deliveries since the 1990s, the M400 (or any other Moller product) has never entered production. After a bankruptcy, the company claims to have initiated a deal with Chinese investors and has shown mockups of a new design that looks like a flying saucer with fanjets around the perimeter. You probably shouldn't hold your breath on this one, but hopefully someone will run with the design, as it is hands-down the handsomest and most futuristic-looking of the pack.

One area that is more likely to reach fruition is military flying cars. DARPA has invested a substantial sum in the creation of what has been dubbed a 'flying Humvee,' and a number of companies have competed for the awards. A leading contender is the Advanced Tactics Black Knight Transformer, which currently looks like a small cargo container with windows, wheels, and eight helicopter rotors mounted atop it. In principle, it follows other multi-rotor designs (even mimicking quadcopter drone design,) but in tests it has flown like a chopper, driven like an off-road vehicle, and can carry well over 4,000 lbs of cargo. It can fly at many thousands of feet in altitude and—best of all—completely autonomously. One key role identified for the craft is autonomous evacuation of the wounded in impassable terrain—where the unmanned

machine could set down, crawl in, and band whisk the wounded away to a MASH unit. Development has some way to go, but has reached working prototype status and holds promise.

In the future we can expect more designs like this; multi-rotor or multi-ducted-fan units capable of flying autonomously to a remote location, driving where they need to go, doing whatever they need to do, and either driving or flying back to the departure point. Flying cars, long a staple of 'I wish' along with hoverboards and jet packs, may soon be a tangible reality.

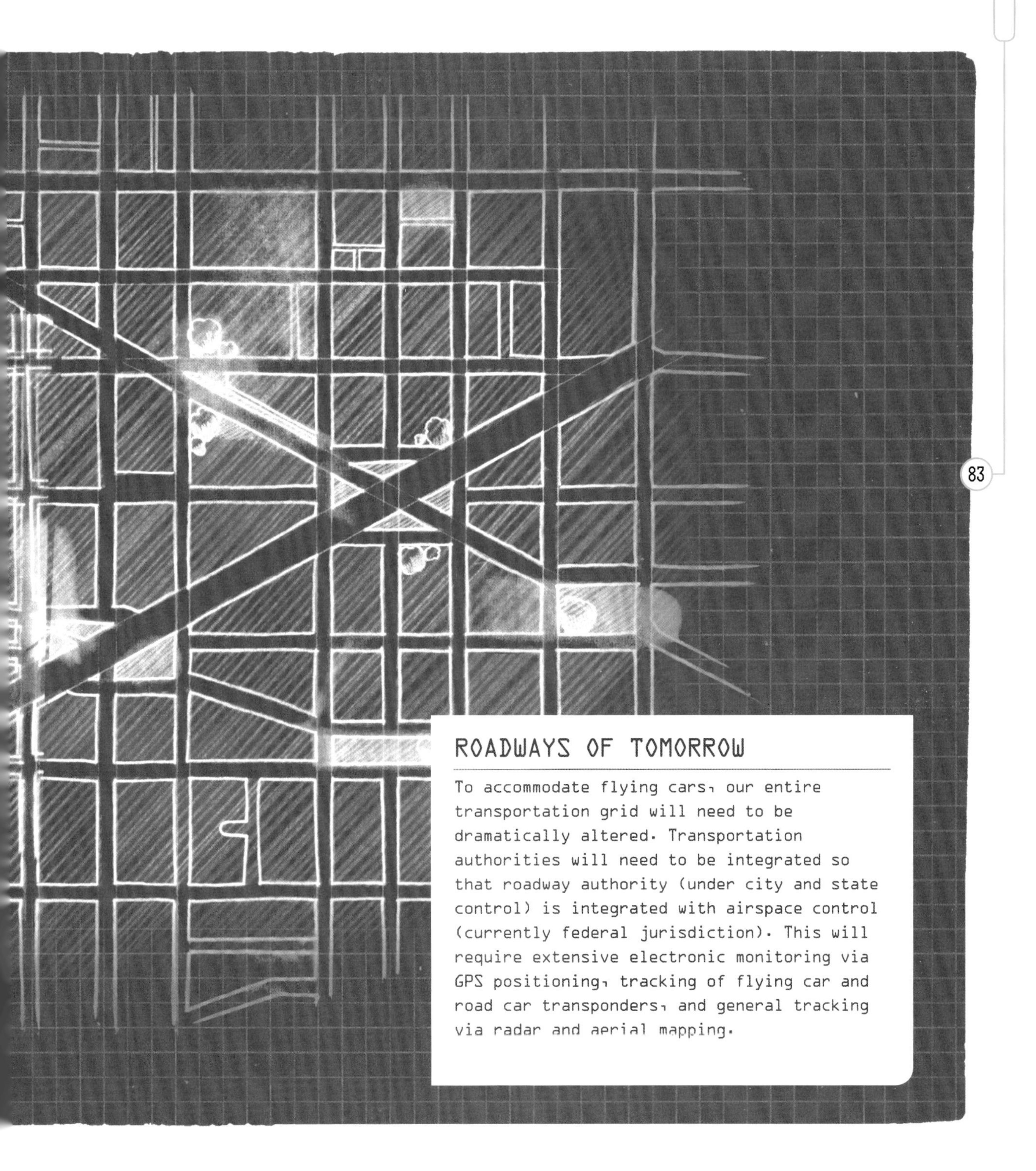

ROADWAYS OF TOMORROW

To accommodate flying cars, our entire transportation grid will need to be dramatically altered. Transportation authorities will need to be integrated so that roadway authority (under city and state control) is integrated with airspace control (currently federal jurisdiction). This will require extensive electronic monitoring via GPS positioning, tracking of flying car and road car transponders, and general tracking via radar and aerial mapping.

LIVING IN SPACE: HOW TO BUILD A COLONY

Ideas about colonies beyond Earth have been around for quite some time. The notion of settling on other planets goes well back into the 20th century; Edgar Rice Burroughs famously placed his John Carter stories (debuting in 1912) on Mars and Ray Bradbury wrote of the planet in quite homey terms in *The Martian Chronicles* (including a fellow named Sam who builds a hot dog stand at a dusty Martian crossroads). But these fanciful tales are about living on other planets—just a substitution of another world for good old terra firma. What about living *in* space? Orbiting habitats have their origins in the early 1900s, when one writer foretold of a 'brick moon,' in which people could live in comfort circling the Earth. By mid-century other futurists were positing more practical metallic structures.

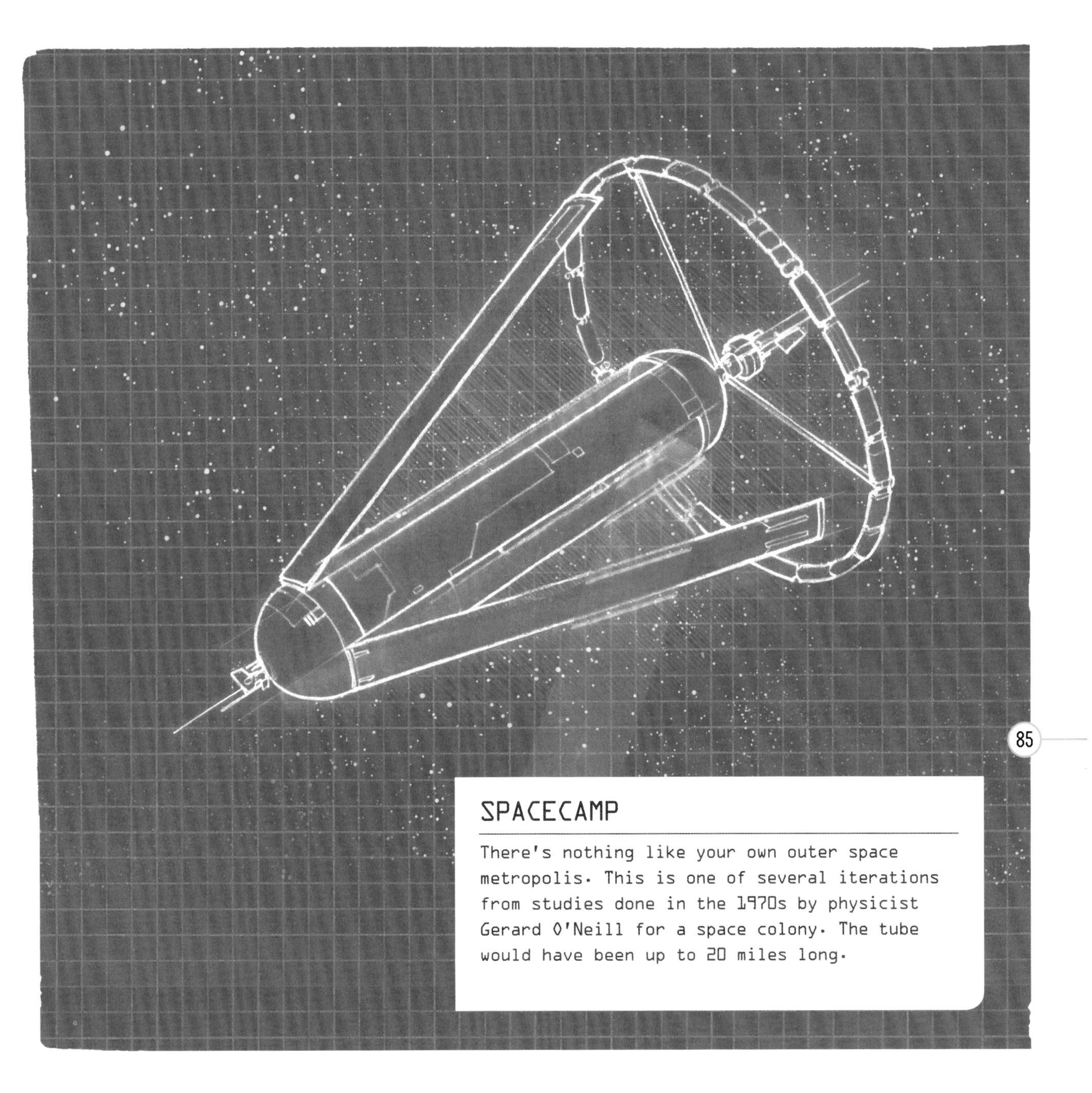

SPACECAMP

There's nothing like your own outer space metropolis. This is one of several iterations from studies done in the 1970s by physicist Gerard O'Neill for a space colony. The tube would have been up to 20 miles long.

EARLY DESIGNS

The whole idea of space colonies was given new life by Gerard K. O'Neill, a visionary scientist at Princeton University. Trained as a physicist, O'Neill began in the 1970s to talk publicly about his ideas regarding the development of space—specifically, people living there. Others toyed with this notion, but it was O'Neill who brought it into the public eye in a dazzling fashion.

O'Neill had developed novel high-energy physics instruments in the 1950s. He also pioneered the mass driver, an electrical catapult designed to launch objects into space without using rockets. In 1977, he founded the Space Studies Institute to provide a forum for his ideas about settling people off Earth. An entire generation of space settlement advocates grew up with O'Neill's ideas guiding their

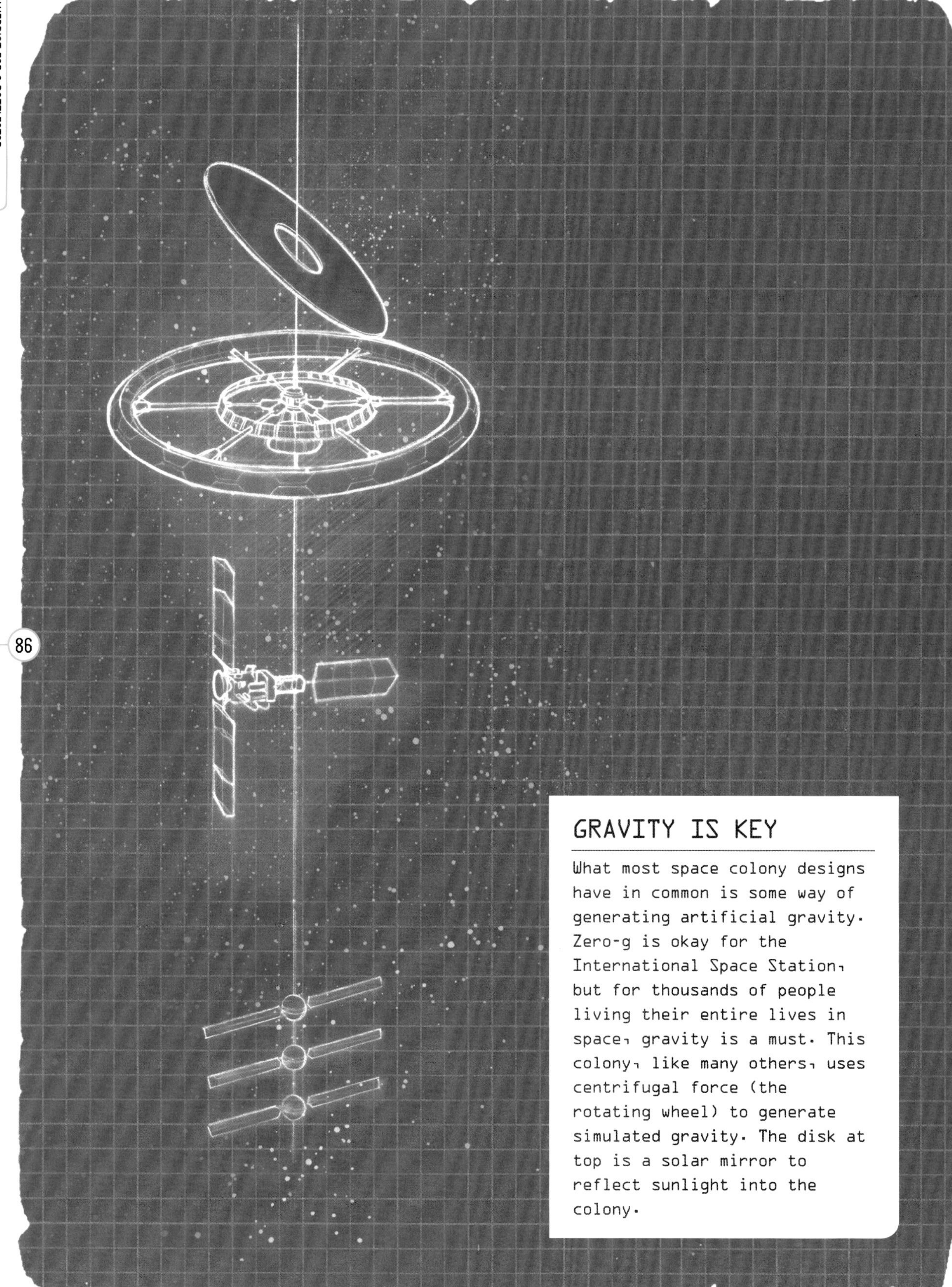

GRAVITY IS KEY

What most space colony designs have in common is some way of generating artificial gravity. Zero-g is okay for the International Space Station, but for thousands of people living their entire lives in space, gravity is a must. This colony, like many others, uses centrifugal force (the rotating wheel) to generate simulated gravity. The disk at top is a solar mirror to reflect sunlight into the colony.

life trajectories, and some of them are now key players in the private spaceflight sector.

Central to his ideas about colonies in space were structures called O'Neill Cylinders. These were massive tubes, some 20 miles long and four miles in diameter, which would be placed in fixed positions in space in pairs. The walls of each cylinder would alternate between solid surfaces and huge windows into space—three of each—with huge mirrors that could be opened and closed to create an artificial day-night cycle adjacent to each of the windows.

The interior of each cylinder would encompass about 500 square miles and a population of several million people. The cylinders would spin along their axis, providing artificial gravity on the inner surface via centrifugal force. Rivers and streams would course through forests created on the inner walls, and there would be sufficient atmospheric volume to allow weather patterns to form: clouds and even rainfall would be part of the weather cycle within.

Another design that emerged about the same time was called the Stanford Torus. This was first proposed at a workshop held at Stanford University in 1975 in coordination with NASA. Similar in concept to the O'Neill design, the primary difference was one of configuration—this one was ring-shaped instead of a tube—and size, with the torus being only a mile in diameter. The interior was a scaled-down version of O'Neill's cylinder, which would accommodate about 10,000 people. Larger versions would hold up to 140,000. The ring would rotate once per minute to create artificial gravity for the residents.

A third design category is that of a sphere in space, called a Bernal Sphere. This was the oldest of the suggested layouts, having been suggested by a British scientist, John Desmond Bernal, in 1929. The ball-shaped main structure would be about ten miles in diameter, and the interior walls would accommodate between 20,000–30,000 colonists. Windows would be at the poles on either end, which was also the the center of rotation that would provide the needed artificial gravity. Again, mirrors reflecting sunlight would simulate day/night cycles. A smaller variety of this structure was suggested by O'Neill in the mid-70s as a more achievable version of his cylinders.

Living in any of these craft would be a unique experience. In the cylinder, you would have an elongated view along its long axis, with the end of the tube vanishing into atmospheric haze and clouds. In the Stanford design, the long view would curve up gradually at both ends and more rapidly to the sides. In the Bernal Sphere, *everything* around you would curve up—you would be living in a bowl. But humans are versatile creatures, and would surely adapt over time. All three designs had areas where zero-g could be experienced; in the case of O'Neill's cylinder, he postulated that people could fly with small wings in the weightless airspace at its center. There's your jet pack at last...except, no fuel required; you can fly all day.

Actually building any of these designs and the dozens of others that have been studied over the years would present substantial challenges. The first is economic: they would be vastly expensive. The International Space Station, which is as close as we have come to something like a space habitat, is just a fraction of the size (about the length of a football field) and costs about $140 billion. As the most expensive structure ever created, it can house up to 12 people, but is optimized for six. It also took 36 flights of the space shuttle, as well as a few Soyuz launches from Russia, to get the required hardware into space.

O'Neill ultimately acknowledged the improbability of constructing his colonies via traditional rocket launches, and went on to suggest using another one of his inventions: a mass driver based on the Moon to launch materials mined and processed there for the assembly of the station. This would have lowered the cost dramatically, but of course itself requires a substantial infrastructure on the Moon. So far, we have had just 12 people visit there, and all they have left behind are footprints, some cameras and backpacks, and the descent stage of their landers (and, for later missions, lunar rovers). We have a long way to go in either scenario.

Another question is where to put space colonies. Low Earth orbit is where the International Space Station lives, but staying in that orbit requires periodic boosts. It is also a place that is filled with bits and pieces of old rocket junk, and at orbital speeds even something as small as a fleck of paint acts like a bullet and is very dangerous to a large structure. So O'Neill and his contemporaries suggested placing their colonies in stable orbits between the Earth and the Moon.

There were other designs, but the description gives you the general idea of what a space colony would look like, and what it might be like to live in one. Recent legislation put before the US congress, H.R. 4257, the SEDS Act (which stands for Space Exploration, Development and Settlement, for you acronym-trackers out there), may help the cause. It is the first piece of legislation put before the US government since NASA was created that would specify, in detail, what their ongoing purpose is. When the agency was created in those crazy space race days, the primary goal was to create a civilian agency to catch up with and surpass the Soviet Union in space, as well as advance scientific and technical accomplishments in space. Since the end of the Apollo program in the 1970s, NASA has ranged far and wide in their efforts to explore space, and each new president seems to have a different agenda, further diffusing what directions are taken. The title page of the bill states this overarching goal:

To require the National Aeronautics and Space Administration to investigate and promote the exploration and development of space leading to human settlements beyond Earth, and for other purposes.

Note the use of the word 'settlements.' The approval of this bill would be a first step towards formally incorporating the idea of off-Earth colonies into NASA's agenda and could result in some ong-term advances towards getting humanity into space—not for short exploratory visits, but to stay.

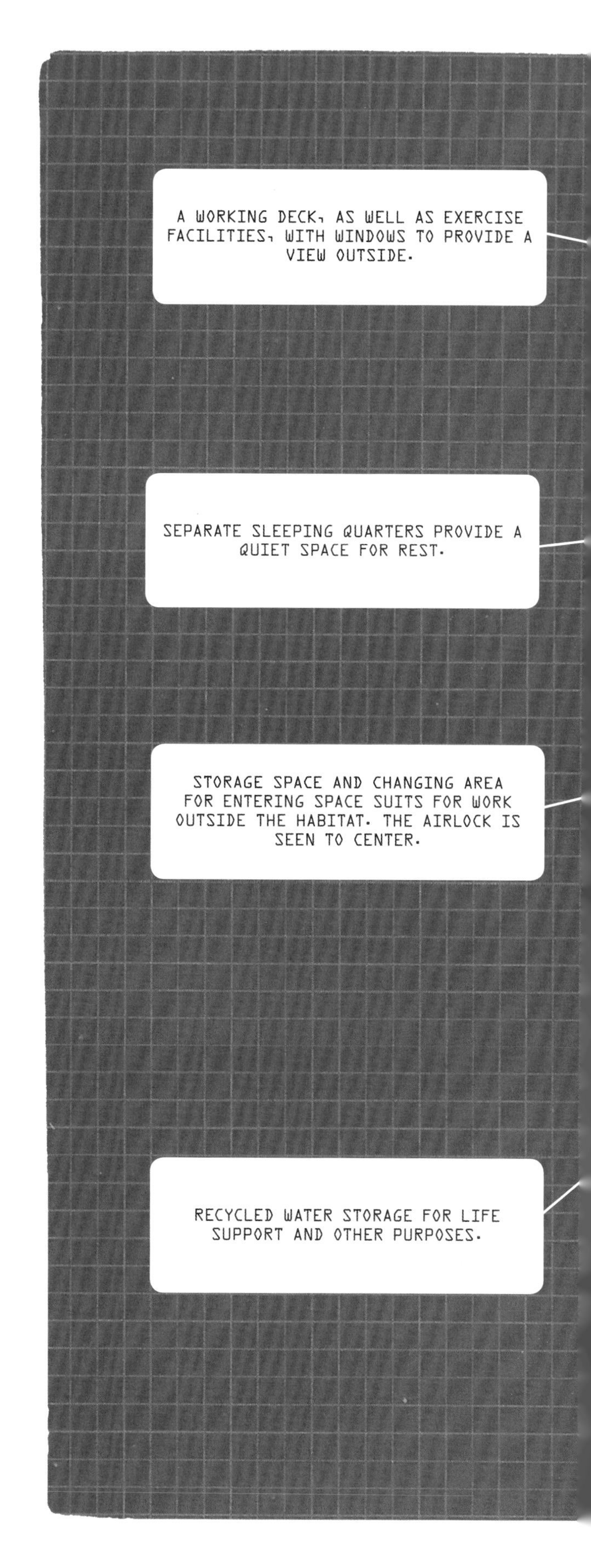

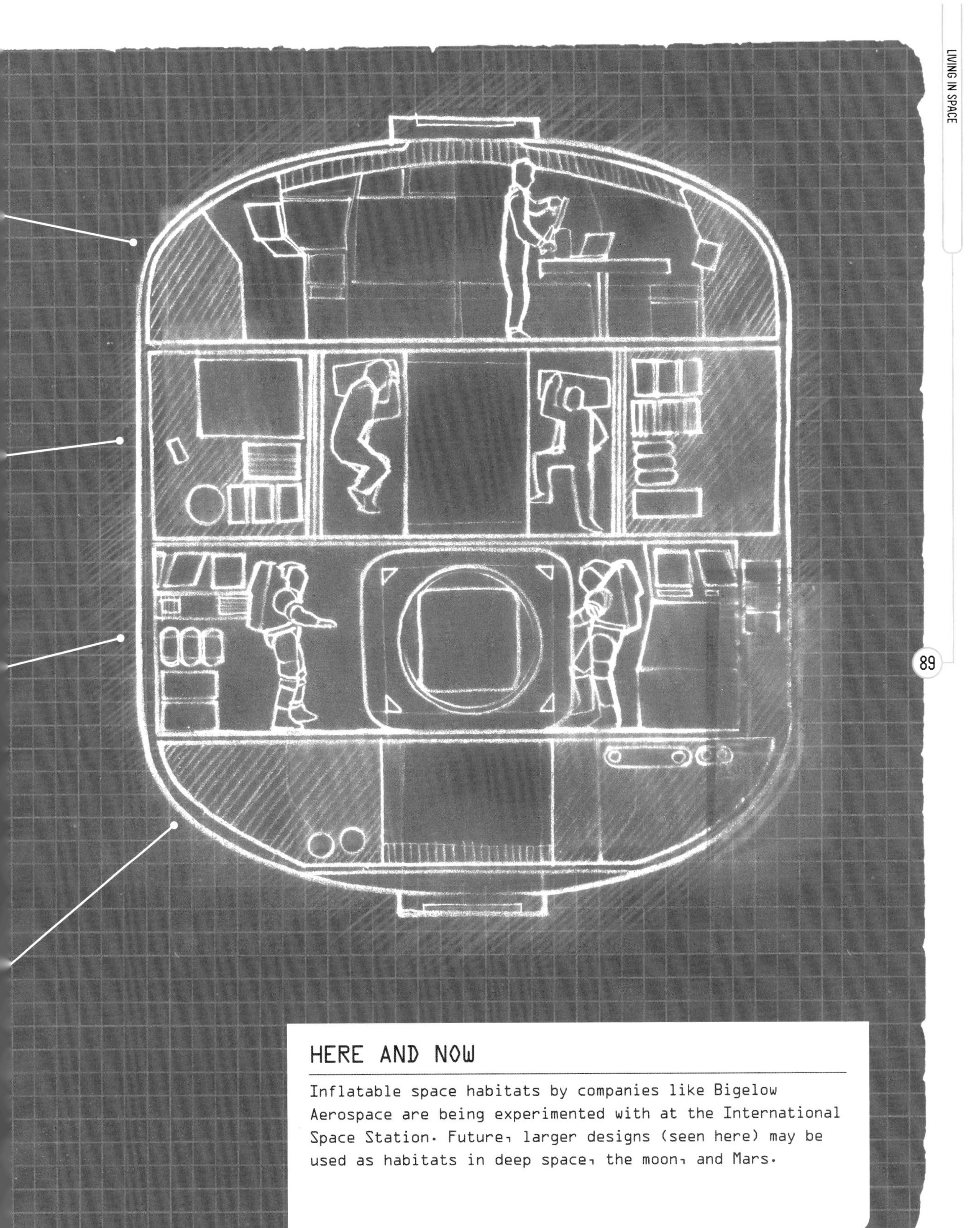

HERE AND NOW

Inflatable space habitats by companies like Bigelow Aerospace are being experimented with at the International Space Station. Future, larger designs (seen here) may be used as habitats in deep space, the moon, and Mars.

WHAT'S NEXT?: CONQUERING MARS

By the time the Apollo program was in mid-stride between 1968 and 1972, plans for 'what's next' were well underway. The lunar program had a goal of reaching, landing on, and exploring the Moon, but the steps to follow were an open question. Should we construct an orbiting space station? Or build a Moon base? Or, should we keep the momentum going and press on for Mars?

THE MARS PROJECT

What actually happened was a bit of the first, none of the second, and absolutely nothing of the third. And while Skylab, America's first space station, was a smashing success in 1973 (as were the Soviet space stations flown at about the same time), it was most certainly not on the scale of a Mars voyage.

It is not as if the interest wasn't there. Both the United States and the Soviet Union had eyes on Mars, though America had an edge, having successfully developed hardware for its lunar missions. The subject of a Mars trek had been well-studied. Wernher von Braun—the German rocket scientist who developed the V-2 rockets, came to America after World War II, and went on to build the Saturn V moon rocket—had plans. As early as 1949, he was writing about designs for a flotilla of spaceships that would fly into Earth orbit and assemble ten larger ships that would then head off to Mars. His epic study was published as a book called *Das Marsprojekt* ('The Mars Project').

Von Braun's fleet would orbit the red planet after a seven-month voyage, and land a small crew of explorers there at the Martian pole. These brave souls would assemble Mars tractors and begin an overland drive to the equatorial regions, where they would build runways to accommodate the other landers/ascent vehicles to take them back to orbit. Of course, this all depended on an enormously

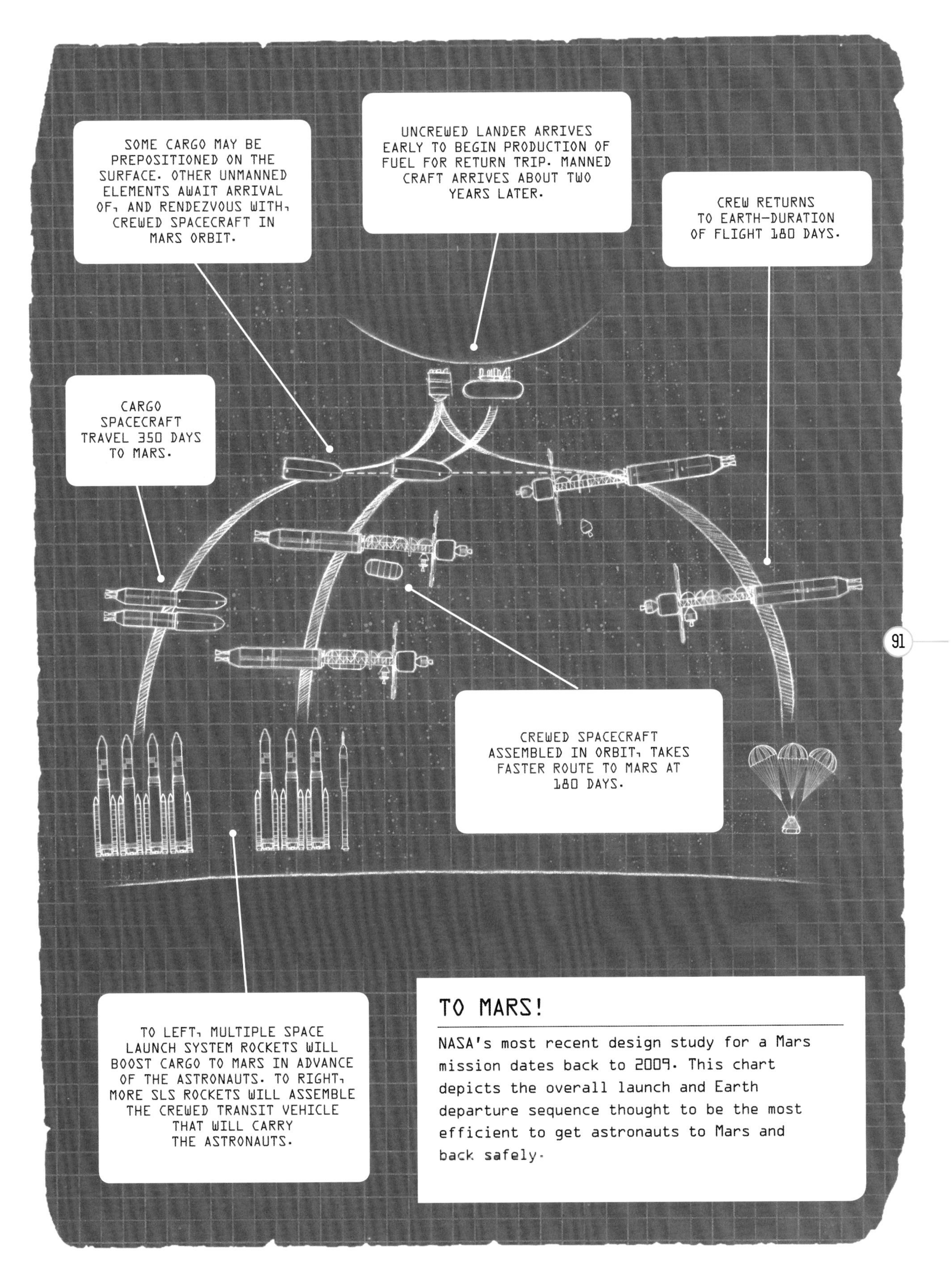

TO MARS!

NASA's most recent design study for a Mars mission dates back to 2009. This chart depicts the overall launch and Earth departure sequence thought to be the most efficient to get astronauts to Mars and back safely.

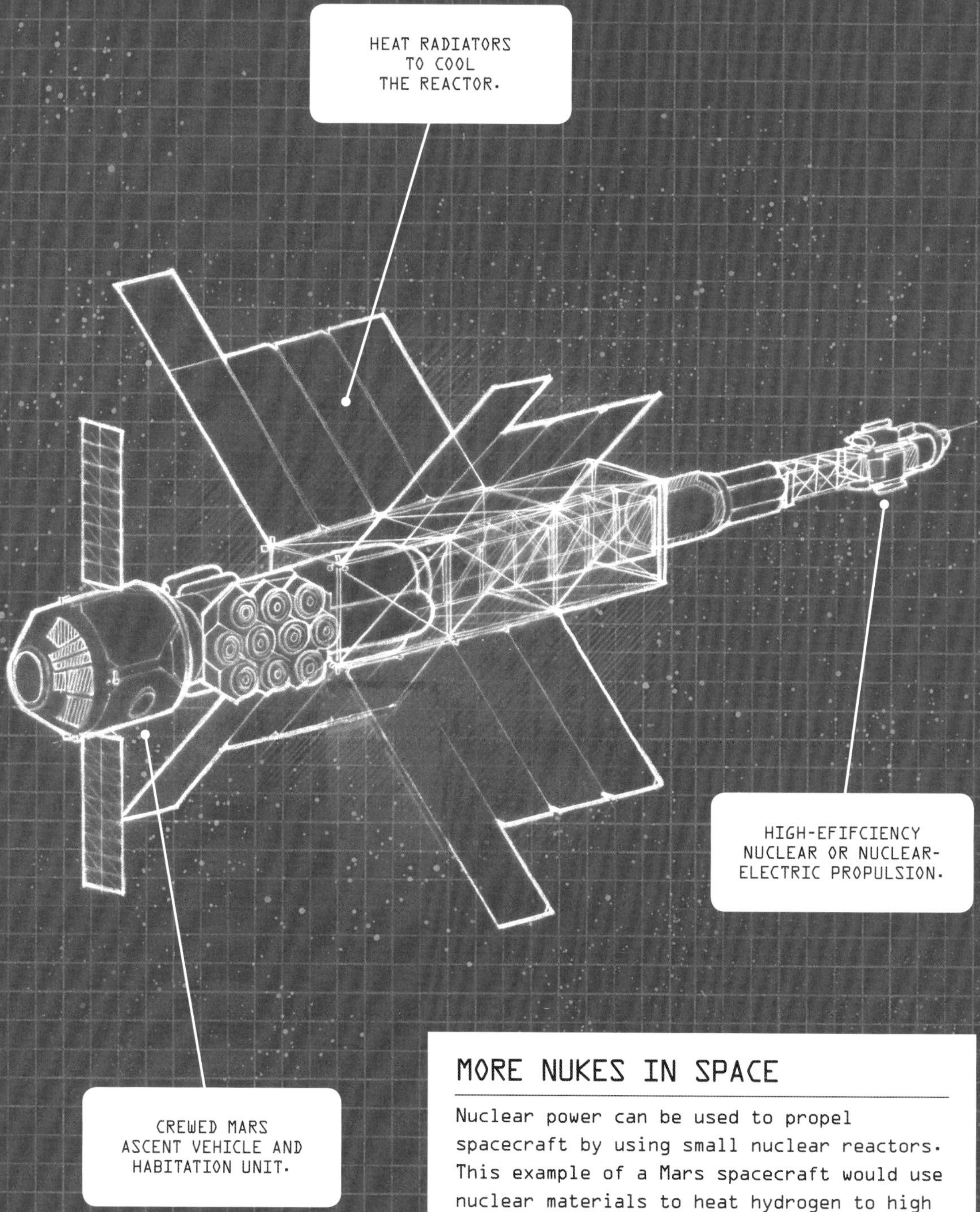

HEAT RADIATORS
TO COOL
THE REACTOR.

HIGH-EFIFCIENCY
NUCLEAR OR NUCLEAR-
ELECTRIC PROPULSION.

CREWED MARS
ASCENT VEHICLE AND
HABITATION UNIT.

MORE NUKES IN SPACE

Nuclear power can be used to propel spacecraft by using small nuclear reactors. This example of a Mars spacecraft would use nuclear materials to heat hydrogen to high exhaust velocity. Such propulsion units hugely decrease the amount of fuel mass the spacecraft needs to carry.

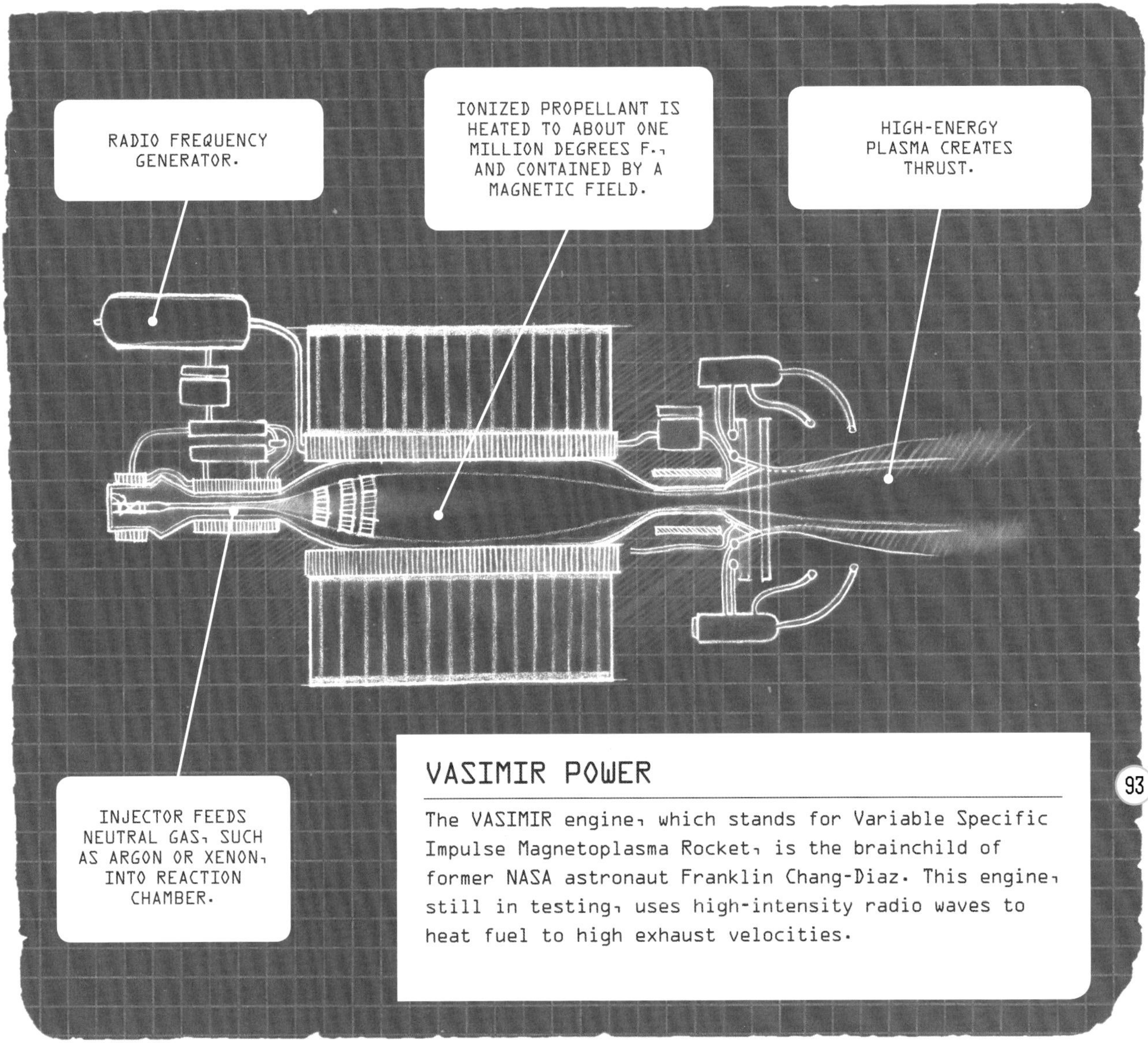

RADIO FREQUENCY GENERATOR.

IONIZED PROPELLANT IS HEATED TO ABOUT ONE MILLION DEGREES F., AND CONTAINED BY A MAGNETIC FIELD.

HIGH-ENERGY PLASMA CREATES THRUST.

INJECTOR FEEDS NEUTRAL GAS, SUCH AS ARGON OR XENON, INTO REACTION CHAMBER.

VASIMIR POWER

The VASIMIR engine, which stands for Variable Specific Impulse Magnetoplasma Rocket, is the brainchild of former NASA astronaut Franklin Chang-Diaz. This engine, still in testing, uses high-intensity radio waves to heat fuel to high exhaust velocities.

expensive and massive program—950 launches would have been required to simply get the raw materials into orbit before departure. It would have been as costly as it was audacious. In the end, aside from some magazine articles, a TV show, and some studies that ended up on NASA shelves, the only crewed Mars missions ever flown were on paper. Boots would not touch Martian soil in the 20th century.

But the major aerospace contractors, hungry to reclaim the hard-earned business of Apollo, even as Moon spending slowed after 1966, ground out study after study. By now von Braun's early work had been set aside—little had been known of the conditions on Mars in 1949. With the US Mariner robotic spacecraft reconnoitering the planet in the 1960s, and the Viking project landing there in the 1970s, we had a much better picture of the challenges Mars presented, and they were vast.

First of all, Mars is far away. At its closest approach, it is still more than 30 million miles from Earth. The Moon, our furthest destination to date, is 240,000 miles away. But it gets worse. You don't fly in a straight line to Mars, you must follow a long, curving path to get there called a Hohmann transfer —it looks just like a chunk of a long, curving orbit. So you are, in effect, aiming for where Mars *will be* when you get there, which is more like 350 million miles. That can take from 150 to 300 days one way. And even that duration depends on how powerful your rocket is, so with current chemical-powered rocket technology it will mean six to seven months in space. Von Braun knew this, which is one reason he was planning to dispatch a fleet rather than a single craft for the long, arduous trip.

A Mars trip presents another challenge. During the entire trek the crew will be exposed to large amounts of deadly radiation. Most of this comes from the Sun, which is pretty friendly to us when we're on Earth, thanks to our planet's magnetic field and dense atmosphere, which deflects most of it. But in space, without that protection, the spacecraft will be hit with all the human-roasting energy the sun hurls at it. Add to this the radiation coming from outside the solar system, and you'd be dead long before you got to Mars unless your spacecraft is well protected.

Then there are the life-support factors to consider. Short of becoming Terminator-like cyborgs, we need to be kept alive for the long trek to Mars. This requires air to breathe, food to eat, water to drink, waste disposal, medical care, and a fairly benign, Earthlike environment-in-a-can. And then we have to deal with the lack of gravity for those many months as well, which does nasty things to human bodies over time. This is even before considering the psychological factors—imagine six months in an elevator with a few other people tossed in to make it interesting.

Some of these factors were known when the Mars mission plans of the 1970s and 1980s were being drafted, but some (like the long-term effects of zero-g) became better understood only as astronauts and cosmonauts spent longer periods of time in space. Since then, crewed Mars mission plans have become ever more involved and expensive. It just isn't as simple as it once appeared.

Ironically, as the years have gone by, the long shadow of the successful Apollo lunar landing program has made plans for a fly-and-come-home mission to Mars a more difficult sell. The 'flags-and-footprints' ethos of Apollo—beat the Soviet Union to the Moon at any cost, then come home, not to return —is a non-starter, at least in the US. The $20 billion it cost to make the Moon voyage, while it developed many of the technologies needed to go further, is simply too much for a single economy to bear in the 21st century (when it equates to more like $100 billion in adjusted dollars.) What is needed, and what should have probably happened on the Moon, is a long-term vision of a lasting presence. This costs more upfront, but would make the venture vastly more rewarding in the long run. The grand scheme, as it is now envisioned, includes using the Moon as a way-station and some plans even suggest using asteroids and the Moon for supplies—the water found on them can provide the raw materials for fuel, water, and even air to breathe.

It will all be expensive and commitment to such a venture is hard to come by. The United States, Russia, and Europe talk about it a lot, but while pieces of the needed tech are being worked on as budgets allow, no large-scale funding has yet been committed to the venture.

China has entered the game in a slightly different vein. That country came to the human spaceflight arena later than the others, sending its first person into space in 2003. The Chinese are now on the cusp of launching their second space station, and have announced plans for a crewed lunar landing sometime around 2020. Mars is on their agenda for 2040+. NASA has said they plan to arrive at Mars sometime around 2035. The private US company SpaceX recently announced plans to send people to Mars, possibly in the 2020s. Who will arrive first? Hard to say.

While NASA has the experience and the basic technology, it is hampered by political considerations. The promise in 1980 that the shuttle would fly once per week was not realized and George W. Bush's plan to return astronauts to the moon was shelved. This is the case for a half-dozen other large initiatives. It isn't NASA's fault; they are subject to Congressional appropriations and suffer further changes of direction every few years as new presidents take office, bringing with them new agendas. The plans may be there, and they have been solid (if sometimes slightly unrealistic,) but there is just not enough support, financial, or political. The Chinese, on the other hand, usually

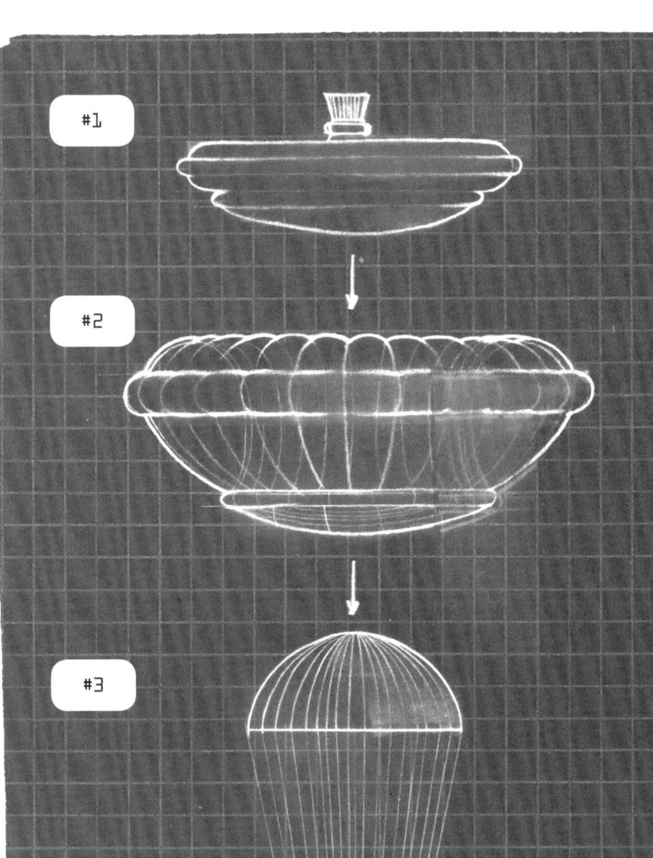

#1 DURING ATMOSPHERIC ENTRY AT MARS, A HEAT SHIELD PROTECTS THE SPACECRAFT DURING THE EARLY PHASE.

#2 WHEN EXTREME HEATING ABATES, A FABRIC RING QUICKLY INFLATES AROUND THE SPACECRAFT, CREATING VASTLY INCREASED DRAG.

#3 WHEN THE LANDER IS SUFFICIENTLY SLOWED FOR SAFE DEPLOYMENT, A PARACHUTE IS EJECTED.

#4 THE SPACECRAFT LANDS, ASSISTED BY BRAKING ROCKETS.

HOW TO LAND ON MARS

The Curiosity rover, at 2000 lbs, is about as heavy as JPL says it can land with current technology. Bigger rockets would help, as would alternative landing technologies such as this: the Low Density Supersonic Decelerator (LDSD), an inflatable donut that slows the lander.

make a plan and stick with it. In the end, the best course may be an international effort, but other than the International Space Station, there's little sign of that materializing.

In the current era, NASA's budget is flat at about $18–19 billion per year. The space station continues to absorb a large part of this, but increasingly money is being steered to the Orion spacecraft and the Space Launch System, or SLS, NASA's new superbooster. These two vehicles, along with SpaceX's Dragon capsule, Boeing's CST 100 spacecraft and the venerable Russian Soyuz capsule, are the near-term future of human spaceflight. Of these NASA's is the only one rated for deep space missions beyond Earth's orbit.

After the cancellation of Bush's Moon program, NASA settled on a program called the Flexible Path. This utilizes the Orion and the SLS to do some or all of the following with human crews: a rendezvous with an asteroid; a station in cis-lunar space (the Earth-Moon neighborhood); a Mars flyby; a landing on a Martian moon; a landing on Mars.

As you can see, the plan is *very* flexible. The only part of this we have committed to at this time is the asteroid rendezvous, which is intended to test and prove technologies for reaching Mars but it appears that the US Congress will scuttle this plan. What ultimately occurs will likely depend on the next presidential administration or the one after.

All this said, the physics of reaching Mars remain the same—it's hard. So how are we likely to accomplish this? Let's look at NASA's flexible approach, since it is the most robust plan to date.

NASA's general plan is to use some combination of Orion and the SLS, along with a habitation module (a version of which is currently under development by a private company.) How this will be powered is not settled, nuclear rockets are a possibility, but the most widely accepted approach utilizes good old chemical rockets—we've flown them for 55 years and they work.

Life support is being aggressively studied on the space station, and the solutions are within reach. Finding a way around the rigors of zero-g is less obvious, but may involve some kind of centrifuge to create a percentage of Earth's gravity on the spacecraft. Less clear is the path toward radiation protection, which will require shielded spaces within the Mars ship. One particularly innovative approach to radiation protection is a jacketed module that contains water and encircles the spacecraft, and researchers even looked at the possibility of growing large algae mats inside it— both water and algae absorb radiation well, and the algae creates oxygen and is edible.

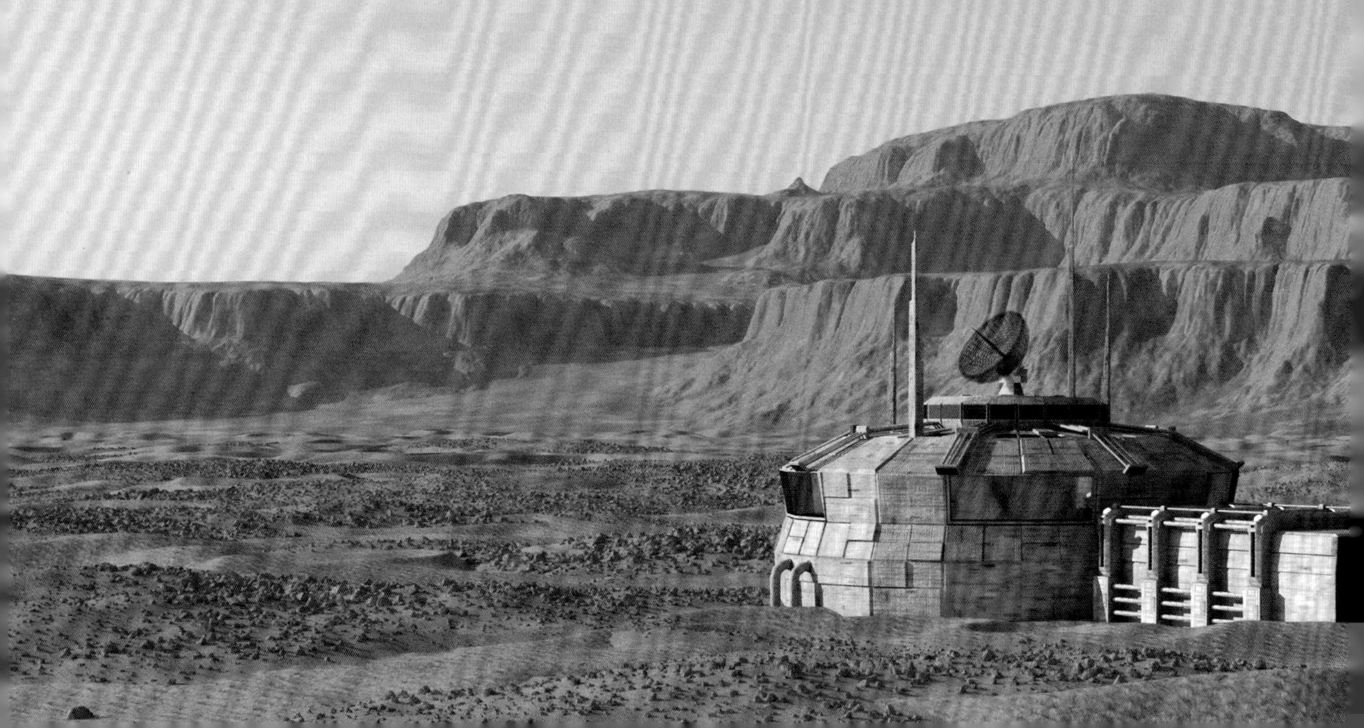

But what about the sustainable infrastructure I alluded to earlier? There are a bunch of ideas, some wilder than others, to grapple with this problem. One thing they all have in common is expense.

Apollo 11 moonwalker Buzz Aldrin has an audacious design that he has been pushing for years. The Aldrin Cycler is a large spacecraft that would be built and set in orbit between Earth and Mars; it circles permanently in a path between the two worlds. Shuttle craft akin to augmented Orion capsules would ferry crews from Earth to the cycler and drop them off for the ride to Mars. Once the cycler reaches the Martian neighborhood, another shuttle takes the crew to Phobos (a Martian moon) or lands on the Martian surface. This would allow for an ongoing exchange of both crews and cargo between Earth and the red planet. It's a smart plan; one that could only come from someone well-versed in the arcane science of orbital rendezvous (Aldrin got his PhD in the subject,) and may offer the best long-term solution to a thorny problem. Other similar proposals have suggested using an asteroid with a sub-surface crew quarters as a cycler, which would deal with the radiation problem.

At the other extreme, a number of one-way plans have been advanced over the years. The most promising is that proposed by Robert Zubrin, a co-founder of the Mars society and an affordable spaceflight proponent. Under Zubrin's plan, unmanned spacecraft are sent to Mars to set up plants to extract water and other usable substances from Martian soil, which can then be used to make much-needed things like oxygen and rocket fuel, as well as drinking water. A crew is then dispatched in a spacecraft that includes a lander, and once they arrive at Mars they are there to stay for at least 18 months. This saves a vast amount of money, since much less mass is being sent aloft—the heavy fuel required for the return trip could be manufactured on Mars. The logic is sound and the designs robust: it just requires a bit more dedication on the part of the astronauts—it's a long mission.

Parts of Zubrin's plan have been wrapped into NASA's current designs, in particular the idea of sending robotic ships to Mars in advance of crews, capable of utilizing Martian resources to reduce the mass of the expeditionary spacecraft that will carry the crew. NASA's Flexible Path states a landing date sometime 'around 2035.' This is generally the same timeframe we have heard for a long time—20–30 years. But this one has some teeth, albeit being dependent on continued funding. Whether or not we make it is more a matter of willpower than engineering.

LIFE—BUT NOT AS WE KNOW IT

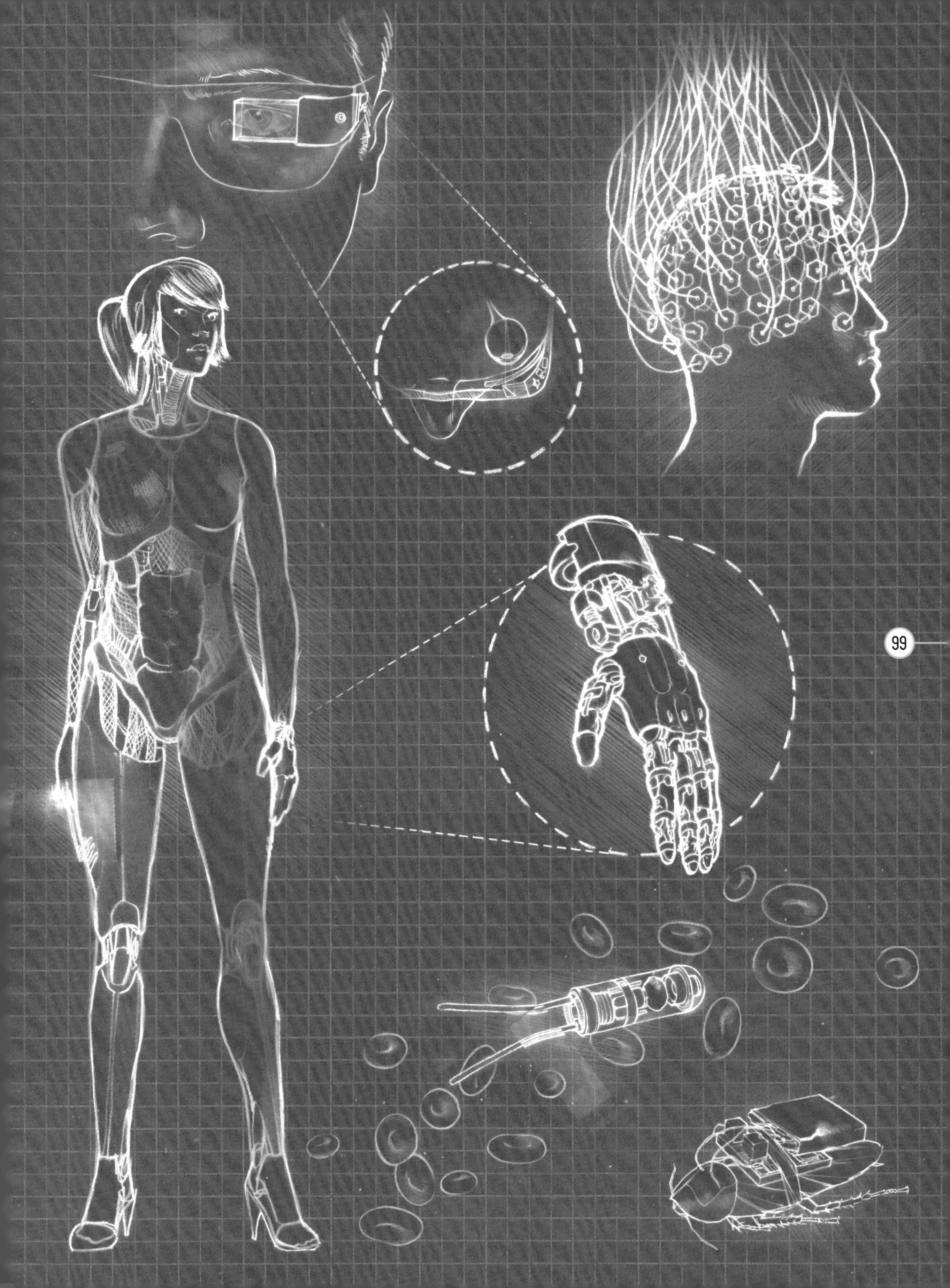

MORE THAN HUMAN: LIVING WITH FEMBOTS

Female robots are a staple of science fiction movies and graphic novels, though almost unremittingly in the form of the shapely mechanical dream woman of the (predominantly male) designers. Robotic females are more properly classified as androids—robots with a human appearance—than as simple robots. A world center in their development has been Japan, where research in universities and government-funded labs has tried to crack the technical challenges of a lifelike, but robotic, woman. After decades of intensive development, they are still very limited in function.

OUT OF REACH?

It has turned out to be much, much harder to create the robots than the examples in books and on-screen made us think. The idea of a talking, walking, and thinking machine seems so simple... but the further along the technology gets, the more the technical challenges of providing such needs as power-packs, realistic skin, and true artificial intelligence become clear. Robots—at least those of the human-mimicking kind—seem to be right up there with jet packs, flying cars, and personal moon rockets in terms of realization... just one more cool technology forever slightly out of reach.

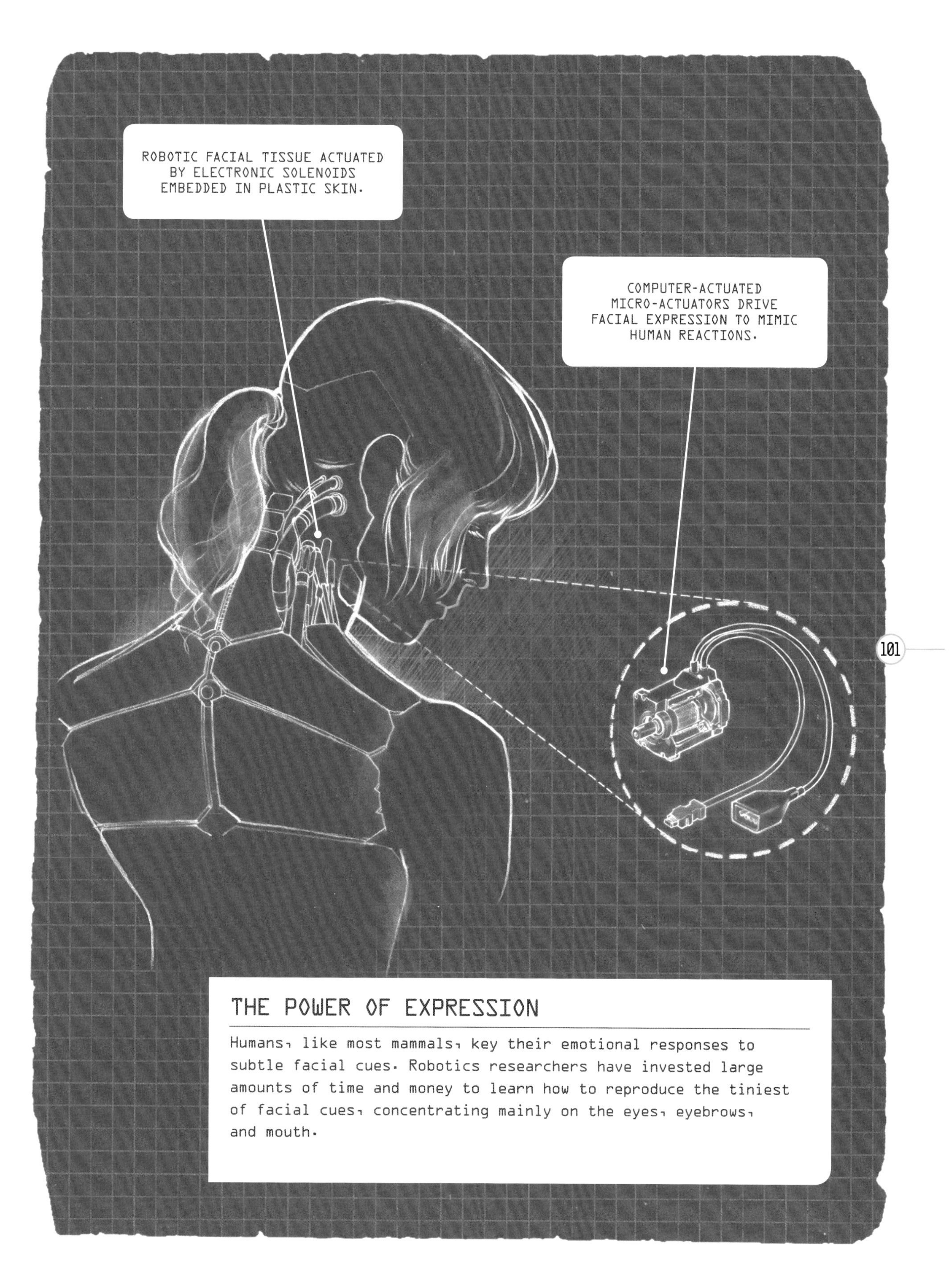

THE POWER OF EXPRESSION

Humans, like most mammals, key their emotional responses to subtle facial cues. Robotics researchers have invested large amounts of time and money to learn how to reproduce the tiniest of facial cues, concentrating mainly on the eyes, eyebrows, and mouth.

Making a machine that can take care of itself and think—ie, understand and reason—is still beyond today's technologies (as engineers on projects such as the Mars Rover and probes to Jupiter and beyond will attest). The Japanese robotics industry has been chipping away at these problems for decades, and while the results are fascinating, practical uses are still at least ten years away.

The first big-screen representationn of a female robot came in 1927 when the visionary German director Fritz Lang released his masterpiece *Metropolis*. In the movie, a mad scientist named Rotwang invents his perfect fembot, a resurrection of the now-dead love of his life. Subsequent on-screen robots, from various Terminators to the Iron Giant, have continued to exhibit a level of intelligence that the real thing is far from matching.

Most robots are program driven, using feedback-loops of some kind to direct their motions. They perform tasks too dull or dangerous for humans to do, so long as the economics of the arrangement are workable. In Japan alone, the number of 'robotic workers' is into the hundreds of thousands (and these are not just a mechanical arm with a welder, but quite sophisticated units), and the country remains in a leading position. Other

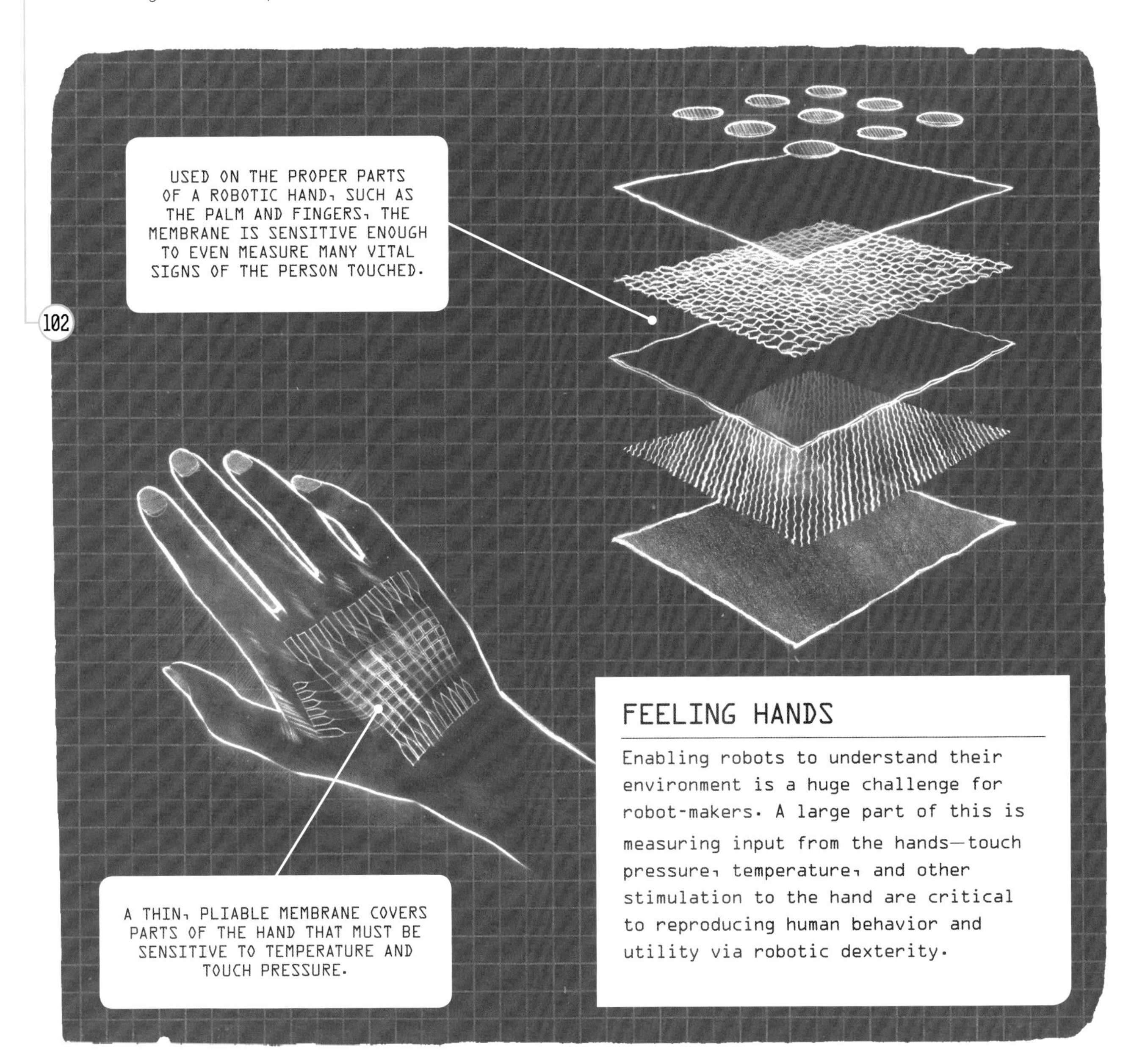

FEELING HANDS

Enabling robots to understand their environment is a huge challenge for robot-makers. A large part of this is measuring input from the hands—touch pressure, temperature, and other stimulation to the hand are critical to reproducing human behavior and utility via robotic dexterity.

developed economies, such as China, Europe, and the United States also lead in the utilization of robotic manufacturing and assembly technologies.

Crude robots have been around for decades, but in 2000 the Honda corporation stunned the world with the introduction of ASIMO (Advanced Step in Innovative Mobility), a walking, vaguely humanoid machine whose primary purpose was to be a technology demonstrator and inspiration to young people. ASIMO can run at almost four mph and stands about five feet tall. While the machine has the ability to walk, run, and even avoid obstacles with minimal human intervention, that is about the scope of its humanoid characteristics. It is also expensive, costing $150,000 a month to lease.

A few years later, another Japanese company began marketing Guard Robo, a mechanized security unit. Able to operate via remote control or autonomously, the machine can follow a programmed patrol path on its own, but any machine-to-human intervention is still overseen by humans, and it is not authorized to shoot autonomously...yet.

More mission-critical is Robonaut, NASA's most advanced robot yet. Robonaut is more humanoid than most, and by far the most anthropomorphic ever sent to space. Designed for dexterity over brute strength, Robonaut could be considered the inverse to Canadarm, the remote robotic arm used on the Space Shuttle. The resemblance to spacesuit-clad astronauts is no coincidence—Robonaut was purpose-designed to work comfortably next to astronauts with the same tools and physical restraints. An improved version, Robonaut 2 (cleverly designated R2) was flown to the International Space Station in 2011 and performed a number of tests successfully during the mission.

All these machines have some level of autonomy. Robonaut 2 was designed to be given specific tasks, then left to finish them without continuous human oversight. A slightly more ominous development would be Atlas, a 330-pound 'robo-sapiens' machine built by Boston Dynamics with money from DARPA. Atlas stands six feet tall and looks an awful lot like the Terminator. At this point, according to Atlas's developers, its mental capabilities amount to roughly those of a one-year-old child. But the machine has been designed with the ability to *learn*. The challenge that DARPA set up to test this semi-autonomous robot involved crossing uneven ground, opening doors, throwing switches, and turning valves.

DARPA has also been busy designing cyborgs—cybernetic organisms—the slightly terrifying blend of people and machines. Early experimentation centered on controlling rodents electronically, but now growing insects with microcontrollers (all the better to spy on the bad guys) is the state of the art. Arming dragonflies with micro-lasers seems to be around the corner.

One of the recent iterations of the Japanese perfection of female androids is called Actroid. First debuted in 2003, Actroid is designed, in the words of its creator, Hiroshi Ishiguro of Osaka University, to be 'a perfect secretary who smiles and flutters her eyelids...' The convergence of humans and machines, and particularly the appearance of part-human, part-machine cyborgs, is raising tricky questions of whether machines can have human rights or at least be protected from sexism.

Other female robots have been created since in the United States and Asia, but Japan always seems to be one mechanical stride ahead of the pack. The developmental emphasis has been primarily in creating realistic facial responses, speech patterns, and overall movement. There has also been work done on sensing, both visual and tactile (yes, the newer models can 'feel' touch, both with their fingers and when you touch them on their plastic bodies).

But there is one huge challenge in creating truly life-like robots: the 'uncanny valley' effect, a term given to it by an early practitioner in the Japanese robotics industry. In short, what we expect to see from another human being, especially in the face, and what we actually see with robots, do not match. The human brain is quite adept at picking up small cues, even if we are not always able to explain them. So when we watch a robot talk or move, our brain says, 'there's something not quite right here...' Crossing the 'uncanny valley'

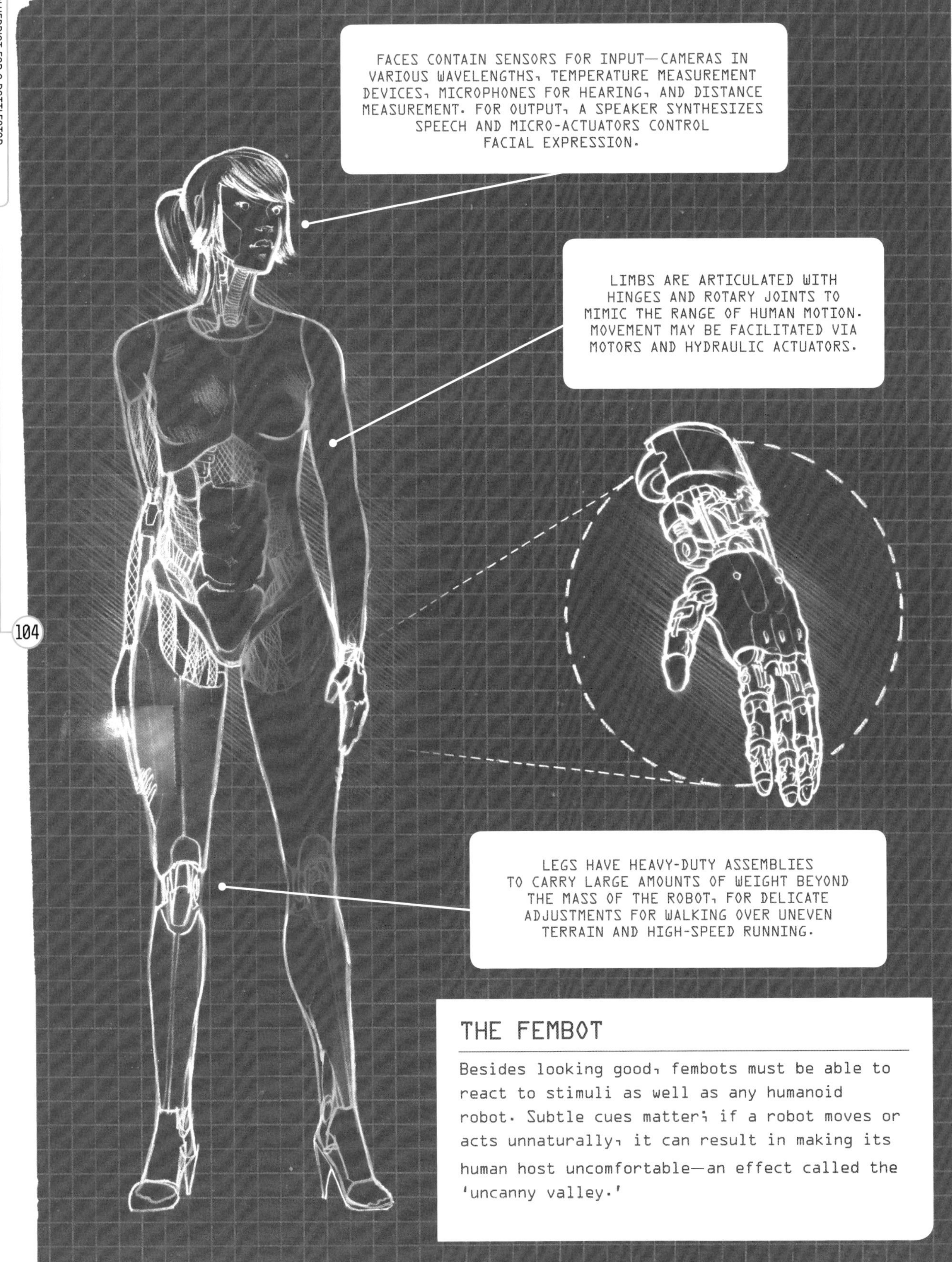

THE FEMBOT

Besides looking good, fembots must be able to react to stimuli as well as any humanoid robot. Subtle cues matter; if a robot moves or acts unnaturally, it can result in making its human host uncomfortable—an effect called the 'uncanny valley.'

has been one of the driving forces behind Japanese robotics.

The answer so far has been largely a mechanical one...dozens of tiny servo-motors drive the facial tissue of the humanized robots in an attempt to mimic human facial cues. As mentioned, this is much harder to do than previously thought. But even more difficult is to program 'smart motivation' behind these cues. To listen to, or visually observe, a person, and then to know how to respond, is a challenge of numbing dimensions for a robotic brain. There have been a number programs that do a convincing job of *pretending* to understand, but true Artificial Intelligence (AI) is still far in the future. When it does arrive, it will be hard in many cases to know if you are interacting with a person or a machine.

This will be the final test of AI, and is known as the Turing test. Turing's hypothesis was that a machine will have passed the test when its responses (as judged by a human being) are indistinguishable from those of another human being. There are many dimensions that this can be applied to—is it simply a text interface, an audio one, or a fully simulated human face talking directly to you? This final category, an artificial face with true AI driving it, will represent the pinnacle of robotics.

Truly independent, mobile, and artificially intelligent robots are probably still a decade or more away, but if designers and engineers like Dr. Ishiguro can get together with DARPA, it could dramatically speed up progress and result in some real-world advantages and progress in robotics to the point where androids could give us humans a run for our money.

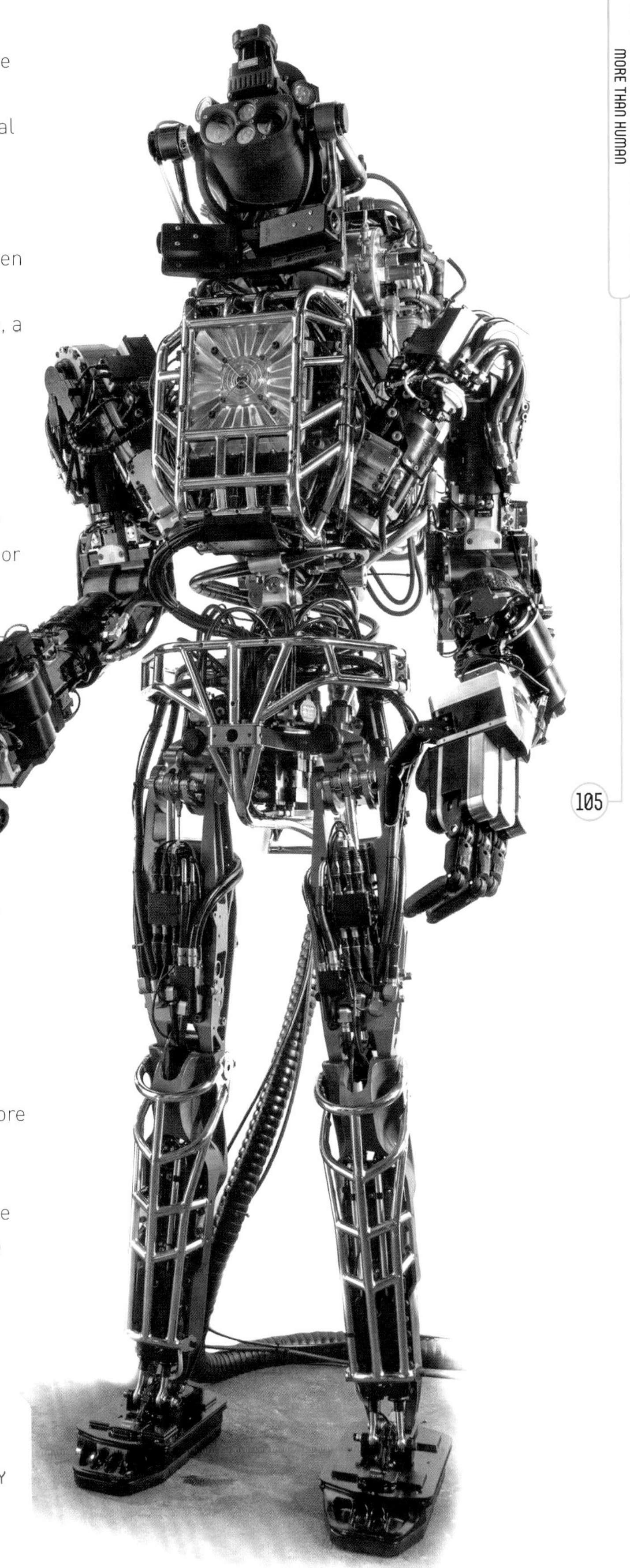

TAKE ME TO YOUR LEADER:
DARPA'S ATLAS ROBOT IS 6 FEET TALL AND WEIGHS MORE THAN 300 POUNDS. WHILE DESIGNED PRIMARILY FOR 'SEARCH AND RESCUE OPERATIONS,' IT SEEMS TO HAVE MILITARY POTENTIAL.

ATOMIC MANUFACTURING: A NANOTECH REVOLUTION

At the far end of the nanotech revolution—the use of miniaturized machines— is a rather disturbing vision of the 'gray goo.' In this extension of the nano-revolution to its extreme, self-replicating machines multiply out of control, devouring everything organic on the planet and leaving only the swarming mass of tiny machines.

NANO-DOCTORS

Nanotech was first mentioned by Richard Feynman of Caltech in a 1959 lecture to the American Physical Society, when he spoke of the possibility of the direct manipulation of atoms. The idea of a swallowable nano-doctor was mentioned. But at the time the technology did not exist to do much of work in the field; the futurists were way ahead of the curve.

Not until the 1980s, when an electron microscope was used to discover fullerenes, did the field begin to grow. A fullerene is a structure made of carbon at the molecular level. The original Fullerenes looked like soccer balls with open sides, but many other shapes were soon engineered, including various tubes and increasingly complex, many-sided

THAT'S TUBULAR

When fullerenes were first discovered, they looked like tiny soccer balls. Soon many other forms were discovered, including this latticed tube, sometimes called a buckytube. Such shapes are incredibly useful for both nano-construction and due to their ability to conduct heat and electricity.

spheres. They were named after the brilliant Buckminster Fuller, an architect, designer, and futurist who brought us all kinds of amazing, geometrically based inventions such as a geodesic dome and oceanic floating cities.

Nanoparticles are tiny. One nanoparticle (which covers a range of objects but is defined as having a diameter less than 100 nanometers) is one-millionth the size of an *ant*. A nanotube is 1/100,000th the width of a human hair. A sheet of typical printer paper is 100,000 nanometers thick. You get the idea.

Even though these particles are vanishingly small, it turns out there are myriad ways of making them useful. Some early research in the 1980s focused on experimenting with the ability to manipulate the basic nano-building blocks into structures, creating novelly nano-items that resembled simple furniture and even tiny replicas of buildings. This soon progressed into making tiny machines: gears, levers, and more elegant devices where one shape moves another just by the effects of drag at such a small scale (for example, one ring inside another is pulled by the

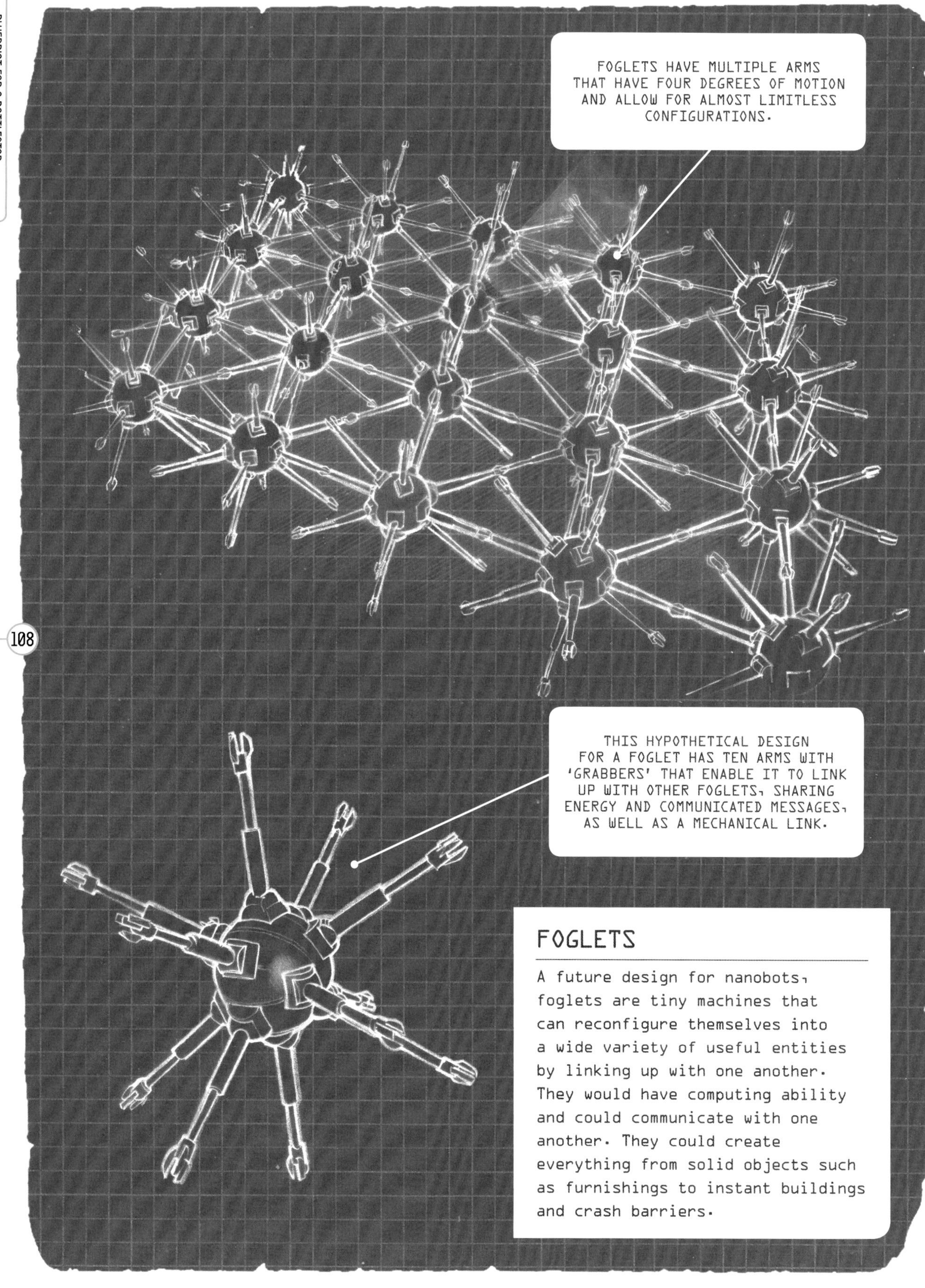

FOGLETS

A future design for nanobots, foglets are tiny machines that can reconfigure themselves into a wide variety of useful entities by linking up with one another. They would have computing ability and could communicate with one another. They could create everything from solid objects such as furnishings to instant buildings and crash barriers.

motion of the first ring). While these simple experiments seem basic, the effects in scale can be impressive.

Other interesting characteristics began to emerge. If a nanotube is placed inside an electrical field, it shoots out electrons like a tiny gun. In the early 2000s, electronics companies like Samsung and Motorola began experimenting with using these electrons to bombard phosphor screens, like old-style TV tubes. In 2010, a Japanese research team mixed nylon with nano-sized carbon fibers and injected this mixture into tiny molds, creating, among other things, gears that could then be used to make atomic-scale nano-machines.

With the current focus on improving humanity's interaction with the environment, other uses were discovered. One example is the hydrogen fuel-cell, which has been in use since the early space race. When hydrogen and oxygen are mixed, electricity and water are generated. But the hydrogen (which can be explosive) must be stored safely. It turns out that hydrogen likes to nestle inside carbon nanotubes, and can be contained within them until needed. The hydrogen is freed when the surrounding temperature and pressure are altered. The benefits for safe storage could be profound.

Another interesting application is the creation of nano-scale artificial muscles to simulate motion. If two carbon nanotubes are joined side-by-side with a thin film in between, and an electrical charge is applied, the tubes will flex, causing the film to do likewise. This mimics the dynamics of human muscles nicely, and with 50 to 100 times the power. The potential applications are huge, in everything from human limb repair to robotics (including fembot facial expressions).

Nanostructures can also be sensitive to certain gases, as their electrical conductivity changes in the presence of these substances. Researchers at the US Naval Research Laboratory were able to detect minute levels of dangerous sarin nerve gas using nanotube gas sensors, and in one-fifteenth the time of existing devices.

There has even been work in the semiconductor area, building nanotubes that act like transistors. The incredible size reduction and the far lower level of energy needed to trigger the on/off state of a nano-transistor could result in yet another momentous reduction in the size of electronic devices.

Nanogears

In 2006, SANDIA National Labs took nanotech a step farther when it created this mechanical gear and catch system at nanometer scale. Since then, nanomachines have become increasingly complex.

But the real breakthrough for nanotech has been in medical applications. Nano-devices are being developed to deliver powerful medications right where and when they are needed most in the human body. Imagine carrying powerful (but often toxic) cancer drugs right to the offending cells, avoiding the dangerous side effects of current therapies.

Fullerenes can transport many therapeutic chemicals, so if the fullerene is attached to an antibody (which already know where to go and what to do), that antibody then finds the diseased entity inside the body, and the medications will leap into action. Proper doses at the proper place—the site of the disease—would massively enhance our ability to treat a wide range of ailments.

Nanomachines with signaling or sensing capability are also valuable additions to the physician's arsenal. C-dots, also known as Cornell dots, are incredibly tiny silica particles that can be infused with a dye that lights up when charged electrically, like gas inside a fluorescent light. Medical experimentation is underway to use these to find cancer tumors, even very tiny ones. This would make the treatment of those growths much easier.

Doctors have for years been performing major surgical work by microsurgery—going in through a tiny incision. However, arthroscopes, the machines that allow the surgeon to see what they're doing in this type of procedure, are still large enough to be a problem—about the size of a fat pencil. Nanotech could revolutionize this, ultimately reducing such devices to the size of a hair and allowing for much less invasive surgeries.

Incorporating nanoparticles—tiny and consistently shaped bits—into many materials can improve the functional properties of items made from these materials. One example is with human bone replacement, a huge issue for people with compromised bone structure. Adding nanoparticles to a polymer, a long chain of synthetic molecules, adds flexing ability and better reactions to compression. Using these materials in bone replacement or enhancement would more closely mimic the ideal structure of bones and result in better lives for millions.

Nanoparticles have even been used to facilitate the 'welding' of flesh. In an experimental setting, gold-coated nanoshells (a tiny core with a thin plating of gold overlay) were added to the site of a cut, and when activated by an infrared laser, they rejoined the two halves of the flesh seamlessly. Being able to do this with wounds (and even arteries and other blood vessels) would be an incredible advance over sutures, which have been the standard approach for a century, or the more modern use of glues.

Research is underway on applications to neurological settings, with various nano-based devices intended to allow computer chips (some of which may eventually be nano-transistorized!) to detect or control neurological functions. Although currently at an embryonic stage, this could eventually range from artificial limb control to treating neurological diseases.

There are plenty of military applications for nano-robotics. Tiny flying drones the size of large flies have been experimented with. In the future swarms of them might fly into enemy territory, sending back information on troop numbers, locations, and potentially even health status (alive, injured, or dead?). Nano-dust has been studied—tiny reactive particles, that when dumped on a battlefield would reveal every contour, construction, combatant, and weapon in the area to overhead scanners in airplanes or satellites.

Taking things even further, another hypothesized use of nanotech would be a system called a utility fog, a vast collection of nanobots that would fill a room. These would be 'universal' nanobots, each having a collection of arms with connectors on the ends that allow for self-assembly, as well as energy and information exchange. When properly programmed, these 'foglets,' as the individual machines are called, would assemble themselves into whatever is required at the moment —they could create create chairs, tables, and even roads, buildings and cities. Foglets could even make air turn instantaneously solid — there's your force-field!. How they would affect respiration and living things is not clear, but the idea is otherwise inspiring (or scary, for those on the wrong end of the technology).

Science fiction has latched on to nanotech in a big way. Authors have called on nanotech to manufacture or 'grow' everything from replacement organs to instant office buildings and space-borne battlecruisers. Much of this probably will be made more efficiently by nanobots one day, but what about the gray goo? While that end-of-the-world scenario is chilling, the likelihood of it occurring anytime soon is slim. It would require the nanomachines to be self-replicating, a technology that is currently quite immature. They could theoretically use substances found in their environment (anything from minerals to organic compounds—like *us*—could be utilized) to reproduce on a vast scale. The resulting devices could be used for all kinds of purposes, from raising buildings to wiping out humanity. The trick, it seems, is to incorporate a set of inviolable rules and controls over the propagation and function of the nano-machines. So when this technology does become available, wise minds will need to labor long and hard to keep it under control.

Some critics of the gray goo scenario have stated that they see the danger of runaway globe-smothering nanobots as unlikely. However, before exploring why, let's look at some of the original assertions. In his seminal book *Engines of Creation*, written back in 1986, engineer and futurist K. Eric Drexler imagined self-replicating nanomachines which—if each could build a copy of itself in 1,000 seconds—would lead to 68 billion of them in ten hours and exceed the mass of the Earth in less than two days.

While this could be used positively, say for planetary engineering, the implications of such technology falling into the hands of terrorists are frightening. It's worth noting that bacteria can accomplish something very similar in terms of self-reproduction, but natural controls and limits to microbial growth have stopped them smothering our planet. In the future, self-reproducing nanomachines will need something similar to keep them in check. Controls could include things such as what they are able to use for raw materials, and limits to the fuels that can provide them with energy. But if some unethical scientist found a way around these limitations, the 'gray goo' scenario might come about after all. So stay alert...

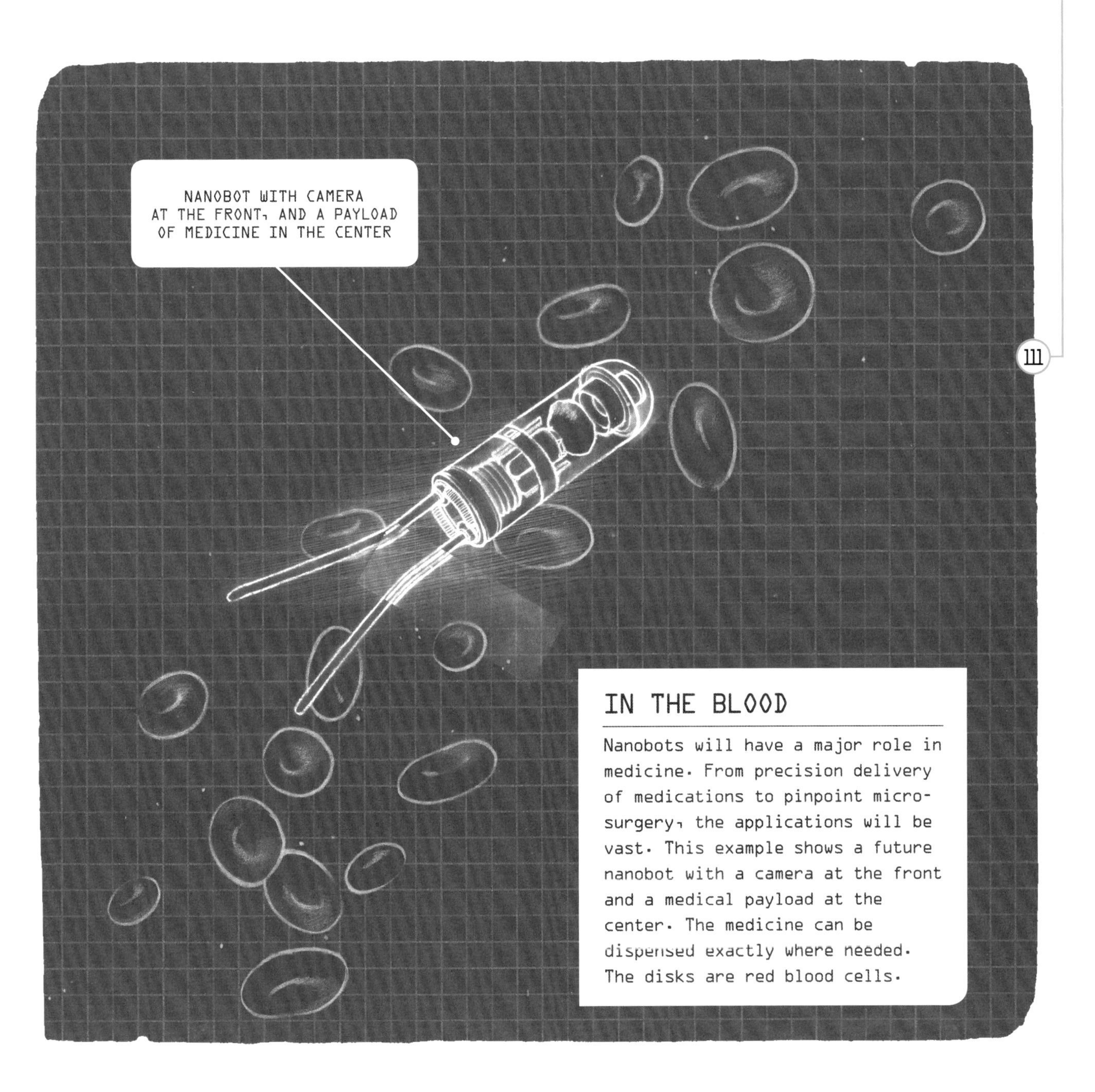

IN THE BLOOD

Nanobots will have a major role in medicine. From precision delivery of medications to pinpoint microsurgery, the applications will be vast. This example shows a future nanobot with a camera at the front and a medical payload at the center. The medicine can be dispensed exactly where needed. The disks are red blood cells.

MAN OR MACHINE?: *TERMINATOR*-STYLE CYBORGS

Like so much of what we are exploring in this book, cyborgs have roots in the past, a presence here and now, and will blossom in the future. But we have already accomplished enough in the field of cybernetic organisms (from which the word 'cyborg' is derived) to know that these bio-machine hybrids are inevitable and, to some extent, already a part of our lives today.

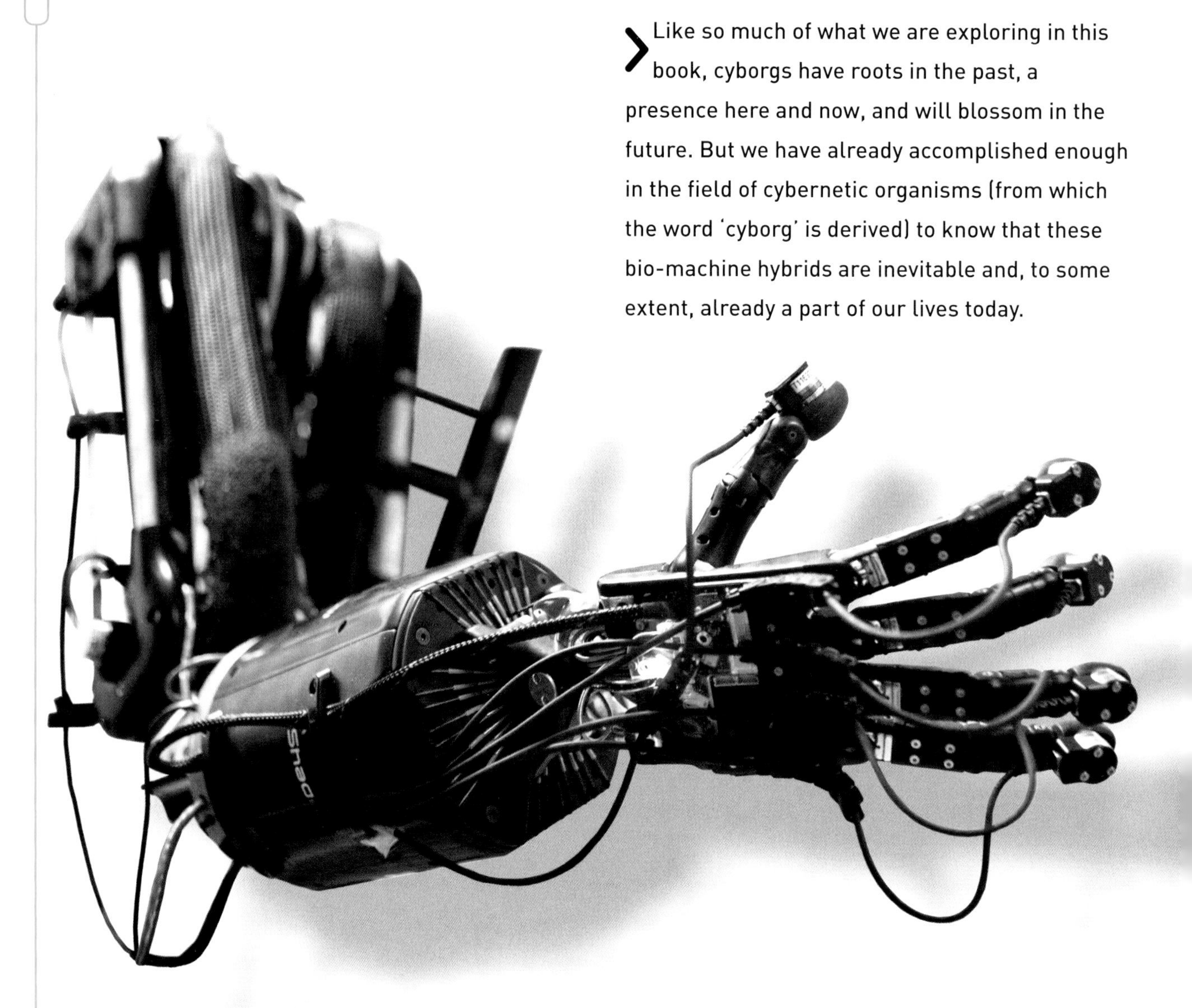

THE BEGINNING

We have all grown up with a general idea of what cyborgs are. The character of Data from *Star Trek: The Next Generation*, the Universal Soldier, the Six Million Dollar Man...these man-machine hybrids seemed like the real deal. And, of course, the Terminator.

The term 'cyborg' was coined in 1960 in an article about creating better astronauts for travel into space. The researchers, Manfred Clynes and Nathan Kline, noted that traditional evolution had altered man to better fit his environment over the years, so,

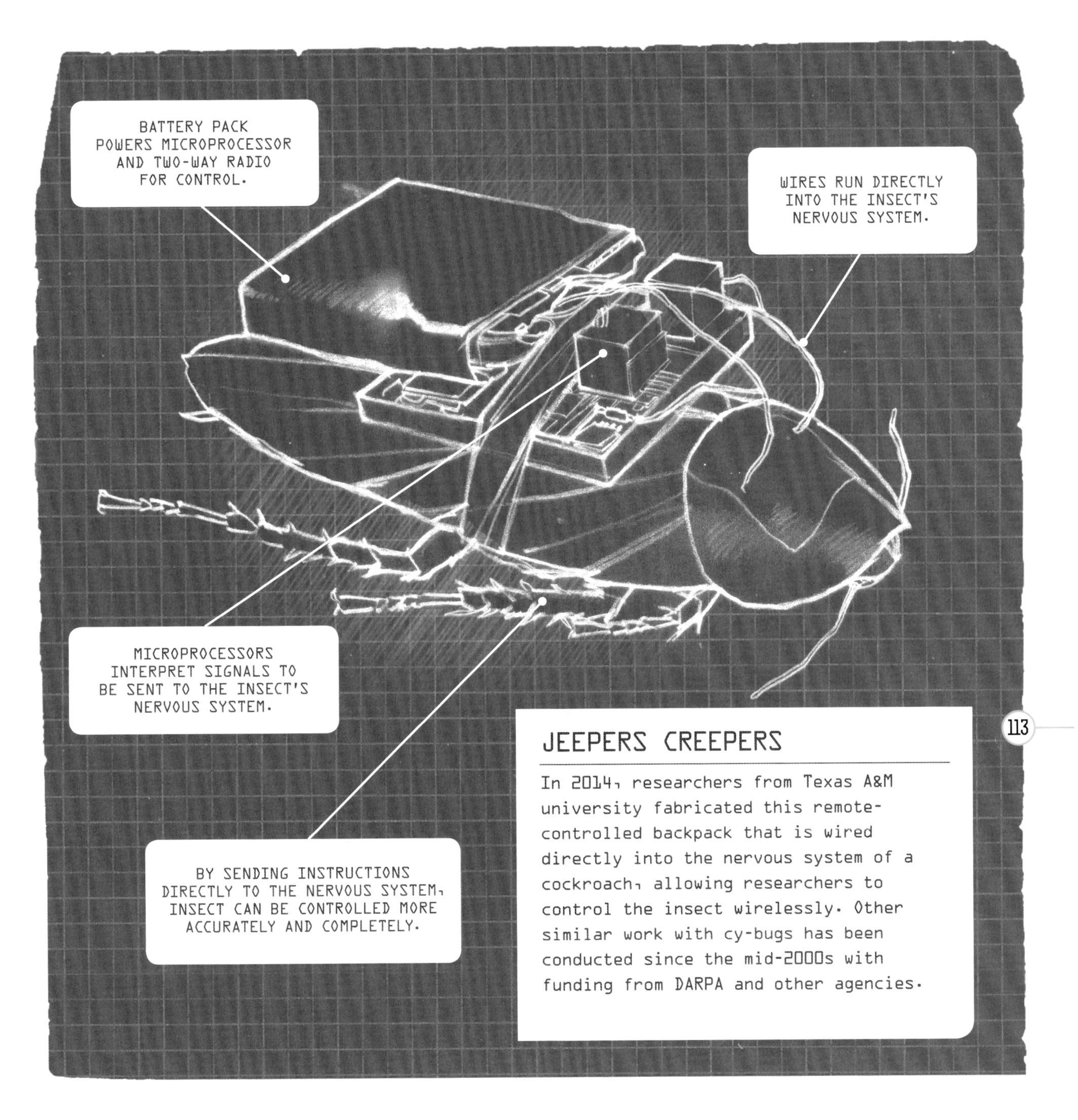

JEEPERS CREEPERS

In 2014, researchers from Texas A&M university fabricated this remote-controlled backpack that is wired directly into the nervous system of a cockroach, allowing researchers to control the insect wirelessly. Other similar work with cy-bugs has been conducted since the mid-2000s with funding from DARPA and other agencies.

they reasoned, why not just speed things up a bit? Using biochemical, psychological, and electronic modifications, anything might be possible. Fortunately, the project did not gain traction.

The idea of improving people via various artificial enhancements had been around much longer—arguably as long ago as Mary Shelly's 1818 novel *Frankenstein*—but the concept of cyborgs has played out in different ways over the years. Enhancement does not necessarily mean that something nasty has to be implanted in you or surgically attached. It can be as benign as Google Glass, which, while not quite the hit some had expected is in a sense a cybernetic enhancement. At its core it is a data-driven extension of the right eye, enhancing our information access and processing.

In the 1960s, Clynes and Kline were suggesting all kinds of interesting enhancements for the human astronaut...artificial systems for handling body fluids, respiration, nutrition, cardiovascular functions, even thought processes and emotions, via the use of tubes, pumps, circuits, and drugs. They even did some experimentation on animals, with a poor lab rat hauling around a cigar-sized osmotic pump attached to its tail to allow the 'continuous injection of chemicals at a slow, controlled rate' without the subject knowing it was being injected.

How about cybernetic organs? Soviet scientists conducted a range of experiments in the 1940s (and probably beyond,) the most disturbing of which involved cutting the heads off dogs and keeping them functioning with a machine that mechanically circulated their oxygenated blood through the brain and head. In some surviving film of the experiment, the dog head can be seen to respond to touch and sound. The poor creature didn't last long.

In 1987, the first permanently implanted artificial heart was successfully tested in a human. Called the Jarvik 7, it was placed into the chest of a man dying of heart failure, who lived 112 days with the little mechanical pump circulating his blood. While the artificial heart had been developed over the course of many years to be the right size, shape, and proper function to preserve blood integrity—blood cells are very fragile—developing mechanical organs has been a tough challenge. Mechanical hearts are still used, but primarily as a bridge to an organic donor heart. Of course, artificial valves and other bits are commonly used today. That first experimental patient, Barney Clark, could arguably be called the first cyborg.

In fact, it could be argued that perhaps ten percent of humanity today are to a degree cybernetic organisms, given the widespread application of pacemakers, an electronic prosthetic, or some other kind of electromechanical augmentation. But these

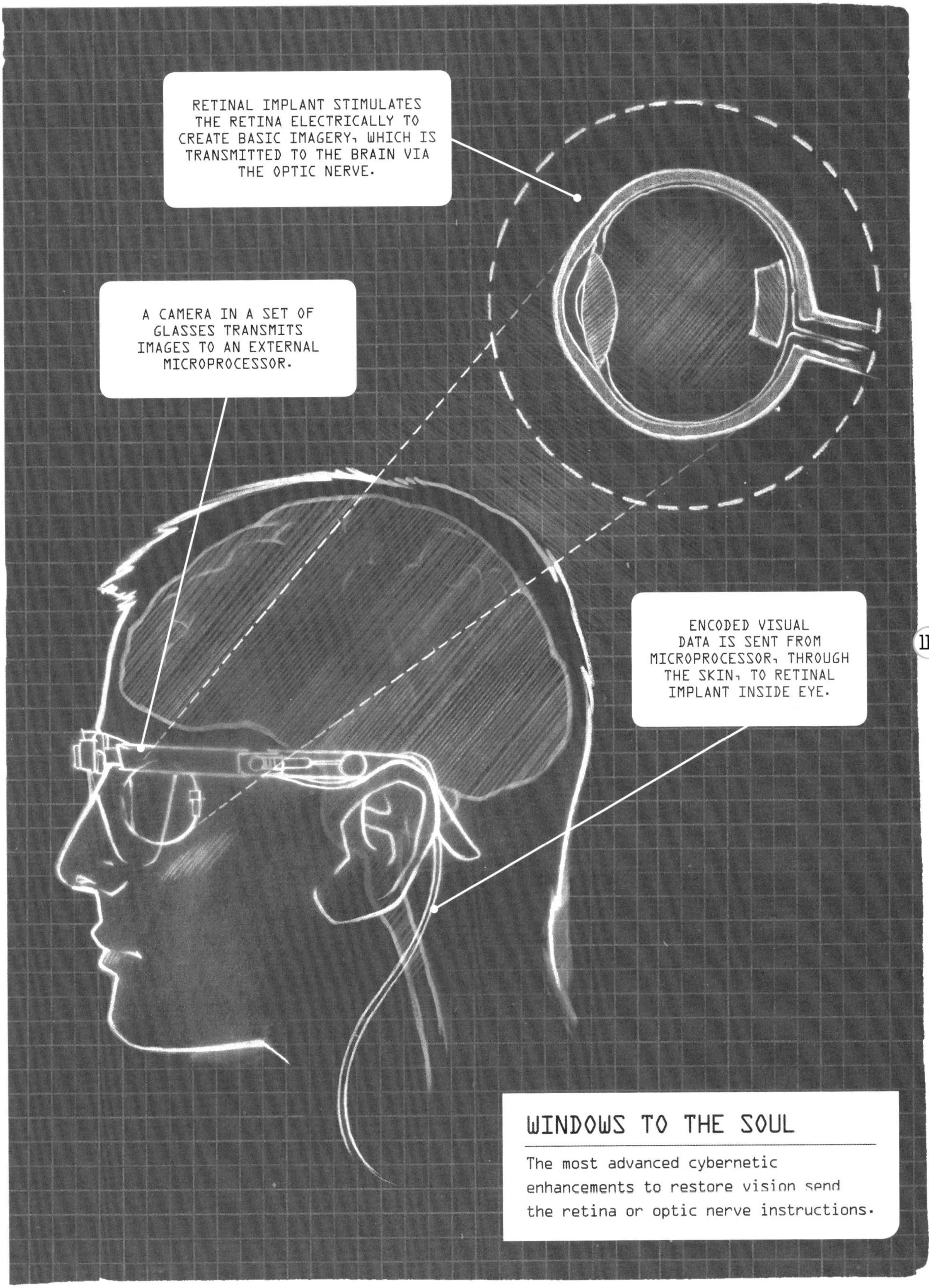

RETINAL IMPLANT STIMULATES THE RETINA ELECTRICALLY TO CREATE BASIC IMAGERY, WHICH IS TRANSMITTED TO THE BRAIN VIA THE OPTIC NERVE.

A CAMERA IN A SET OF GLASSES TRANSMITS IMAGES TO AN EXTERNAL MICROPROCESSOR.

ENCODED VISUAL DATA IS SENT FROM MICROPROCESSOR, THROUGH THE SKIN, TO RETINAL IMPLANT INSIDE EYE.

WINDOWS TO THE SOUL

The most advanced cybernetic enhancements to restore vision send the retina or optic nerve instructions.

hardly count as the kind of Terminator-style being we imagine cyborgs to be.

The most advanced research is being done on non-human subjects. In 2009 came the announcement of insectoid cyborg research that resulted in successful control over the behavior of the subject bugs. The University of Michigan conducted a study in which researchers implanted electrodes into the brains and flight muscles of a beetle. Shocking the optic center made the insect take flight, and it could then be steered via stimulation of its flight muscles. In 2012, a North Carolina State University lab surgically attached a small circuit board to the back of a cockroach with wires attached to its antennae. By sending current to the right and left feeler, they were able to steer the roach. The bug-sized backpack had a battery and radio receiver, so it could be joysticked to do as commanded.

Similar experiments have been done with lab rats at the State University of New York, using transmitters implanted in their brains that enable operators to steer and track them as they sniff out various scents. In this way, gas leaks could be identified, people found within earthquake rubble, and explosives sniffed out.

Mini-Drones

Robotic mini-drones are the future of surveillance in situations where secrecy is required. This artist's impression of a mechanical fly-sized drone is one example of how these tiny audio/video devices can be disguised as everyday insects.

This research is funded by DARPA, and the work is said to be intended to be used for search-and-rescue operations. The small animals and insects can go where humans and dogs cannot, and at substantially less risk. But the military applications are obvious—not only would a cyborg dragonfly or hummingbird be able to penetrate enemy strongholds to send back data, and possibly even video, but there is no reason to assume that they could not be equipped with small syringes filled with a neurotoxin, or some kind of easily spread biological agent.

Even more outlandish, in 2009 a lab at Tokyo University was able to transplant the head of a moth to a small robot. The moth's sense of smell was transmitted to, and controlled, the robotic platform. Chillingly, the moth had been dead from the time the head was transplanted, so in a real sense, this was a zombie cyborg.

The first self-professed human cyborg is Neil Harbisson. Born in 1982 without the ability to detect color (a form of extreme colorblindness), Harbisson took matters into his own hands. In 2004, he had an antenna surgically implanted into the back of his skull, which looks uncannily like a reading lamp for an eBook. The device allows him to hear, inside his brain, tones representing the colors in a given area. He can also connect with the Internet via the device, and receive phone calls directly into his brain.

Robotic limbs are another area of cyborg development, one that could be of great benefit to millions of amputees. Jesse Sullivan lost both his arms in 2001 when he was working for a power company in Tennessee and came into contact with a live 7,400-volt power line. A few years and a reported $6 million later, he has two prosthetic, robotic arms with electrodes wired into his chest. When he grasps or touches an object, signals travel to the nerves in his chest, indicating grip pressure and even temperature. Signals traveling the opposite direction enable him to manipulate his arms. He says that he can do dishes, shave, and drink from a bottle, things

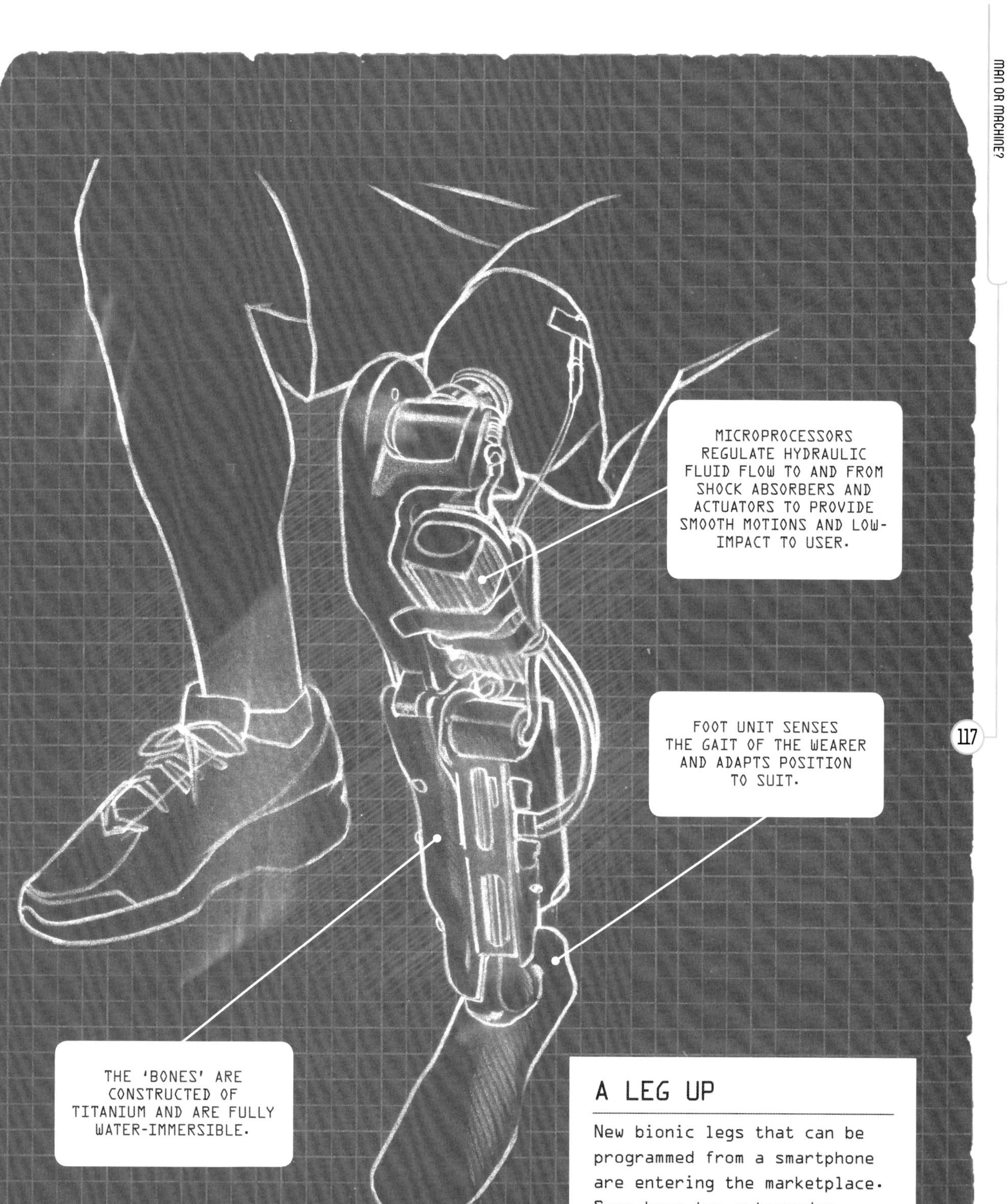

A LEG UP

New bionic legs that can be programmed from a smartphone are entering the marketplace. Some have two auto-modes, which sense the wearer's needs and adapts functions to suit, as well as the capability to accept direct input from the smartphone.

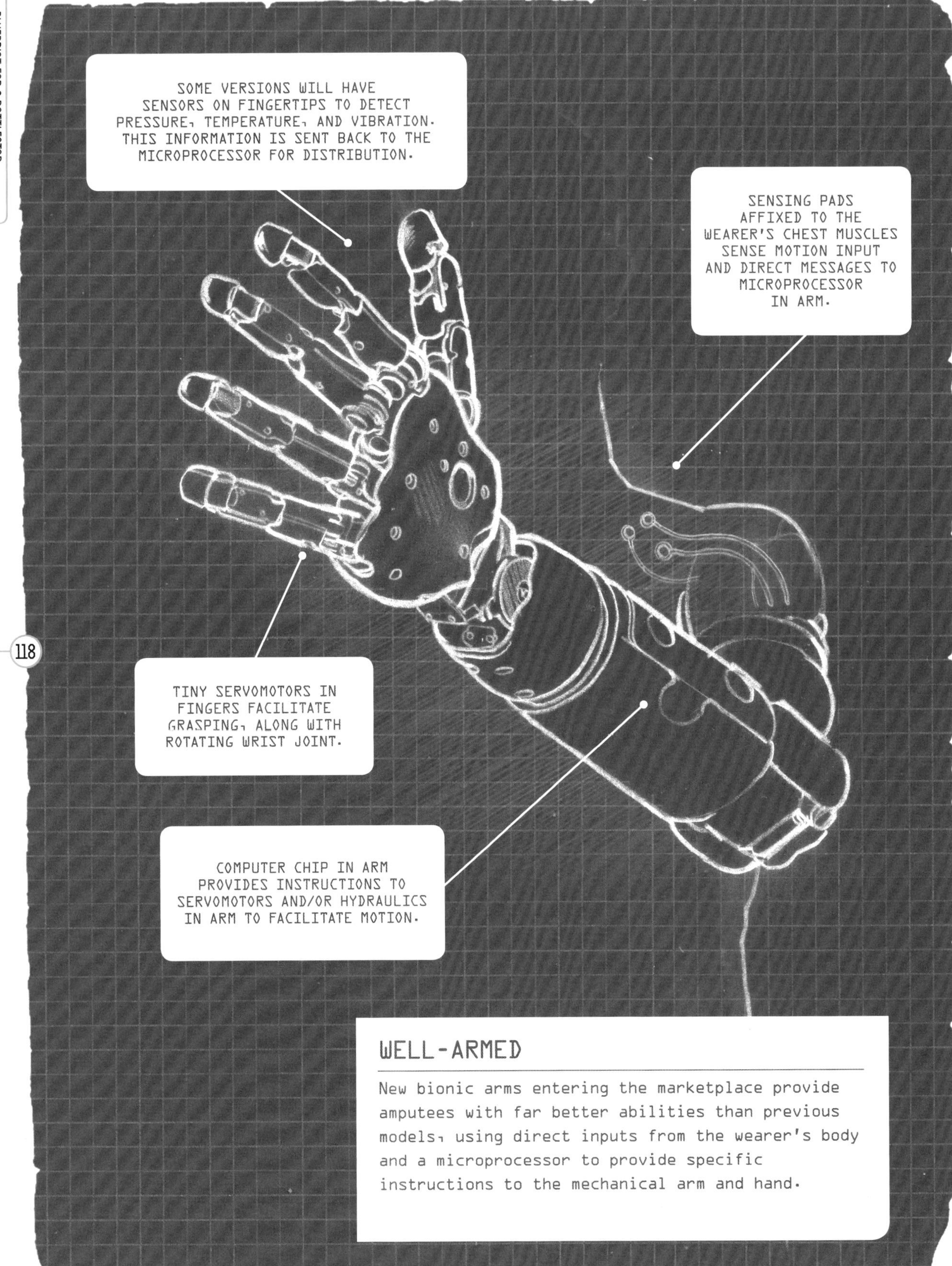

WELL-ARMED

New bionic arms entering the marketplace provide amputees with far better abilities than previous models, using direct inputs from the wearer's body and a microprocessor to provide specific instructions to the mechanical arm and hand.

very difficult to do with standard mechanical prosthetics.

In their new Biological Technologies Office, launched in 2014, DARPA has begun researching various technological enhancements to 'restore and maintain warfighter capabilities.' Some of these 'enhancements' are intended to make injured soldiers whole again, a worthy objective.

One cyber-device being researched is a chip that, when implanted in the brain, will enable soldiers who have suffered severe brain trauma to regain their ability to create new memories, a task that many find challenging or impossible now. The program is also experimenting with brain implants that can control prosthetic limbs, a next-generation solution to Jesse Sullivan's bio-controlled mechanical arms. And a metallic-magnetic additive to human blood is being developed that could be guided to the site of an injury, instantaneously sealing the blood vessel from within.

Yet not all cybernetic enhancements need to be physically attached. Q-Warrior, a British military project, is an outgrowth of Google Glass-type technology which attaches to a helmet and gives soldiers greatly enhanced situational awareness by projecting images over the battlefield. Enemy positions, identity transponders, even combined infrared and satellite images of objects invisible due to darkness or being blocked by walls or foliage become visible. Important data and statistics can be scrolled through, and information on combat-readiness of fellow soldiers can all be made available. The greatest challenge to such a system is the density of information and a human's ability to understand and implement it quickly, but nobody has yet developed a direct neural implant capable of communicating that level of data...at least, not that they will admit to.

Cybernetic Circuitry

This is a false-color scanning electron microscope (SEM) image of a neuron (nerve cell, orange) on a silicon chip. The cell was cultured on the circuit until it formed a network with nearby neurons (for example, far left). Under each cell is a transistor, which can excite the neuron above it. The neuron then passes a signal to the other neurons attached to it, which activate the transistors beneath them. This may be the future of electronic and biological integration, resulting in true cybernetic organisms at the smallest scales.

The Q-Warrior data display will integrate well with the US Army's Tactical Assault Light Operator Suit (or TALOS for short). This exoskeleton combines the most advanced hardware that the military can devise. In essence, it is a robot driven by a biological organism—a human—and hence, a cyborg. When it is eventually deployed, the helmet will have sensors and a Q-Warrior-like tactical display. The suit will enhance physical ability with electronics and motors. Body armor will be incorporated into the exoskeleton. The soldier's weapon may be integrated, displaying remaining ammunition and barrel temperature. There have been prototype self-aiming weapons. Some medical intervention may be built in to deal with injuries. Finally, there has been talk of liquid-impregnated armor that hardens when struck by a projectile—similar technology has been used for years in such mundane application as sports car shock absorbers (hydraulic fluid that thickens when electricity is passed through it).

Incredibly, all this means that in cyborg terms, we really are very close to the Terminator.

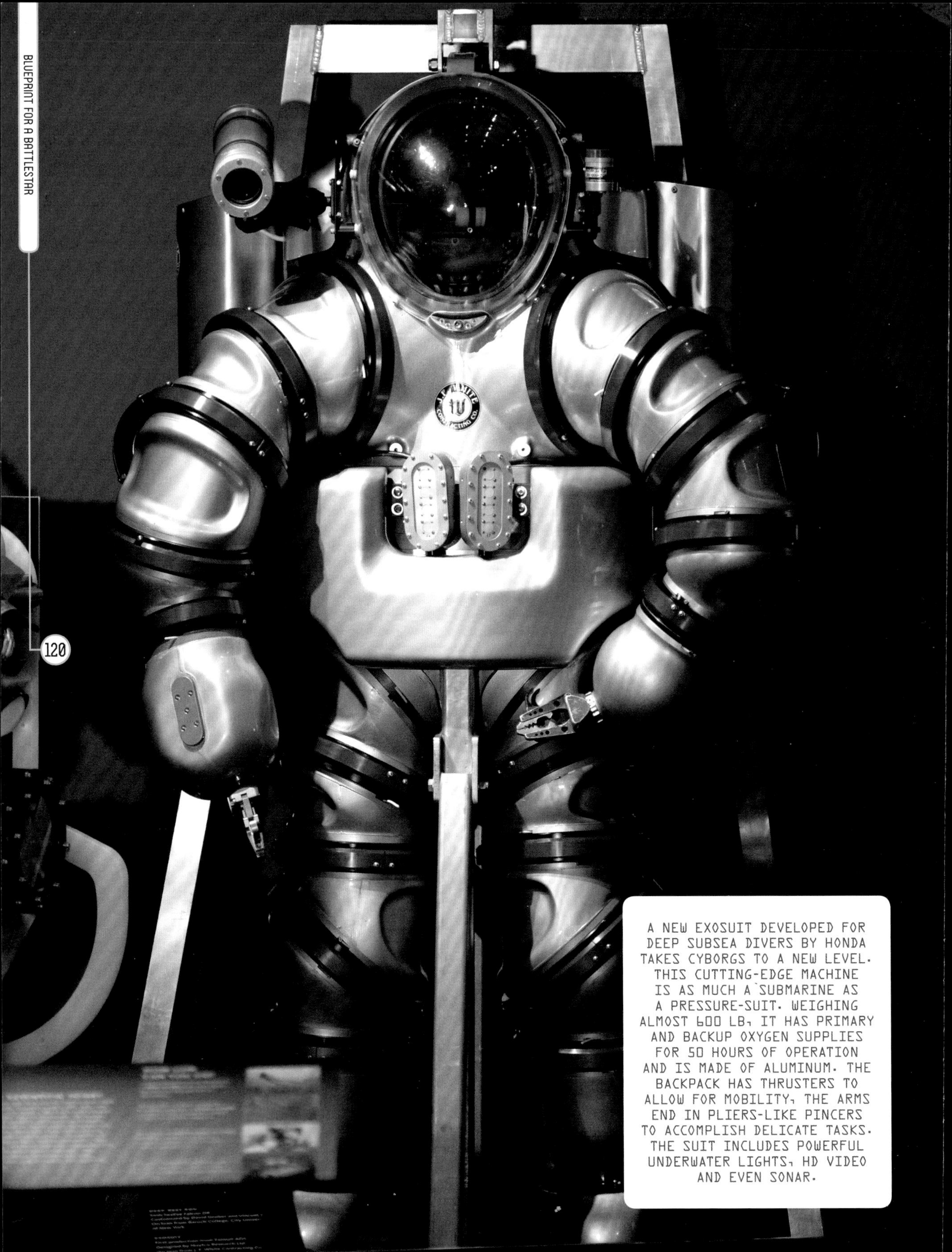

A NEW EXOSUIT DEVELOPED FOR DEEP SUBSEA DIVERS BY HONDA TAKES CYBORGS TO A NEW LEVEL. THIS CUTTING-EDGE MACHINE IS AS MUCH A SUBMARINE AS A PRESSURE-SUIT. WEIGHING ALMOST 600 LB, IT HAS PRIMARY AND BACKUP OXYGEN SUPPLIES FOR 50 HOURS OF OPERATION AND IS MADE OF ALUMINUM. THE BACKPACK HAS THRUSTERS TO ALLOW FOR MOBILITY, THE ARMS END IN PLIERS-LIKE PINCERS TO ACCOMPLISH DELICATE TASKS. THE SUIT INCLUDES POWERFUL UNDERWATER LIGHTS, HD VIDEO AND EVEN SONAR.

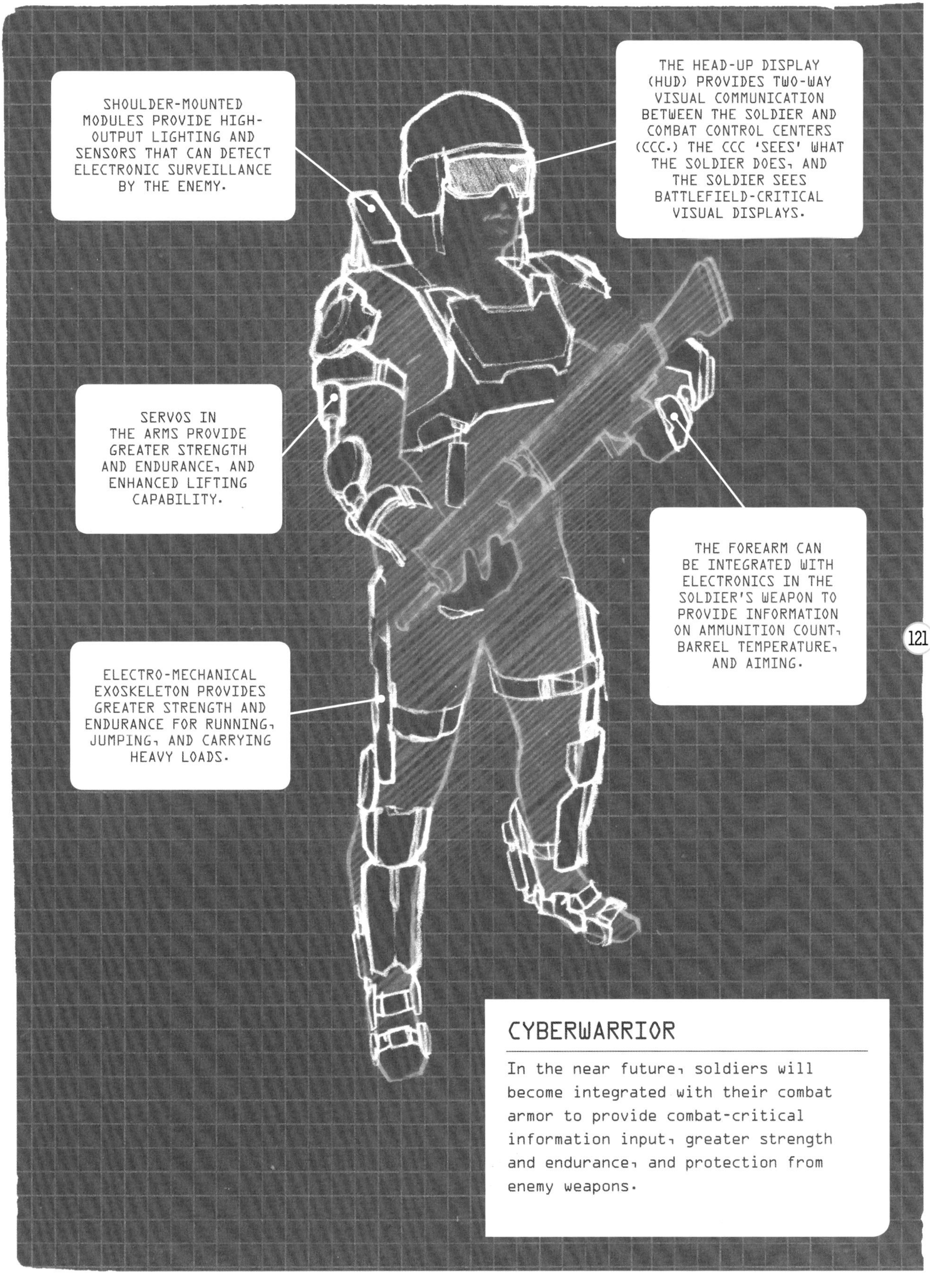

SHOULDER-MOUNTED MODULES PROVIDE HIGH-OUTPUT LIGHTING AND SENSORS THAT CAN DETECT ELECTRONIC SURVEILLANCE BY THE ENEMY.

THE HEAD-UP DISPLAY (HUD) PROVIDES TWO-WAY VISUAL COMMUNICATION BETWEEN THE SOLDIER AND COMBAT CONTROL CENTERS (CCC.) THE CCC 'SEES' WHAT THE SOLDIER DOES, AND THE SOLDIER SEES BATTLEFIELD-CRITICAL VISUAL DISPLAYS.

SERVOS IN THE ARMS PROVIDE GREATER STRENGTH AND ENDURANCE, AND ENHANCED LIFTING CAPABILITY.

THE FOREARM CAN BE INTEGRATED WITH ELECTRONICS IN THE SOLDIER'S WEAPON TO PROVIDE INFORMATION ON AMMUNITION COUNT, BARREL TEMPERATURE, AND AIMING.

ELECTRO-MECHANICAL EXOSKELETON PROVIDES GREATER STRENGTH AND ENDURANCE FOR RUNNING, JUMPING, AND CARRYING HEAVY LOADS.

CYBERWARRIOR

In the near future, soldiers will become integrated with their combat armor to provide combat-critical information input, greater strength and endurance, and protection from enemy weapons.

MY PET *T-REX*: HOW *JURASSIC PARK* GOT IT WRONG

Dinosaur toys are a perennial favorite with young children: little plastic ones, wooden puzzle skeletons, picture books. All those adults who had played with them as five-year-olds had memories of their childhood *T-rex* companions revived when Michael Crichton's *Jurassic Park* came along, first as a novel in 1990 and then as the start of one of the world's most successful film franchises in 1993.

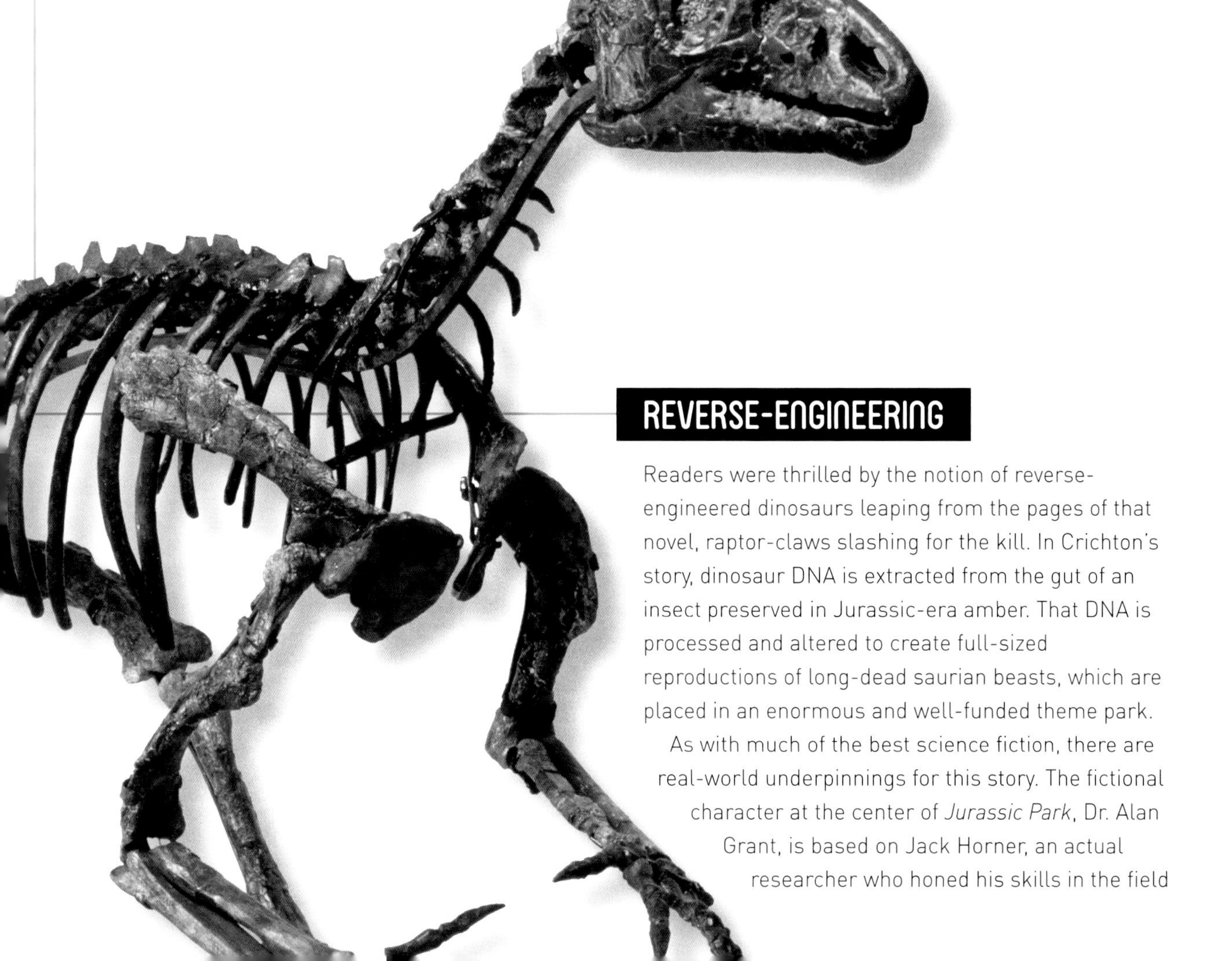

REVERSE-ENGINEERING

Readers were thrilled by the notion of reverse-engineered dinosaurs leaping from the pages of that novel, raptor-claws slashing for the kill. In Crichton's story, dinosaur DNA is extracted from the gut of an insect preserved in Jurassic-era amber. That DNA is processed and altered to create full-sized reproductions of long-dead saurian beasts, which are placed in an enormous and well-funded theme park.

As with much of the best science fiction, there are real-world underpinnings for this story. The fictional character at the center of *Jurassic Park*, Dr. Alan Grant, is based on Jack Horner, an actual researcher who honed his skills in the field

THIS DINOSAUR FOOTPRINT WAS FOUND IN OTJIHAENAMAPARERO, NAMIBIA, AFRICA. AT ONE TIME, A RELATIVE OF MODERN BIRDS LEFT THIS IMPRESSION IN MUD THAT LATER HARDENED INTO ROCK. MODERN GENETIC ENGINEERING MAY BE ABLE TO RECREATE SUCH CREATURES... FROM CHICKENS

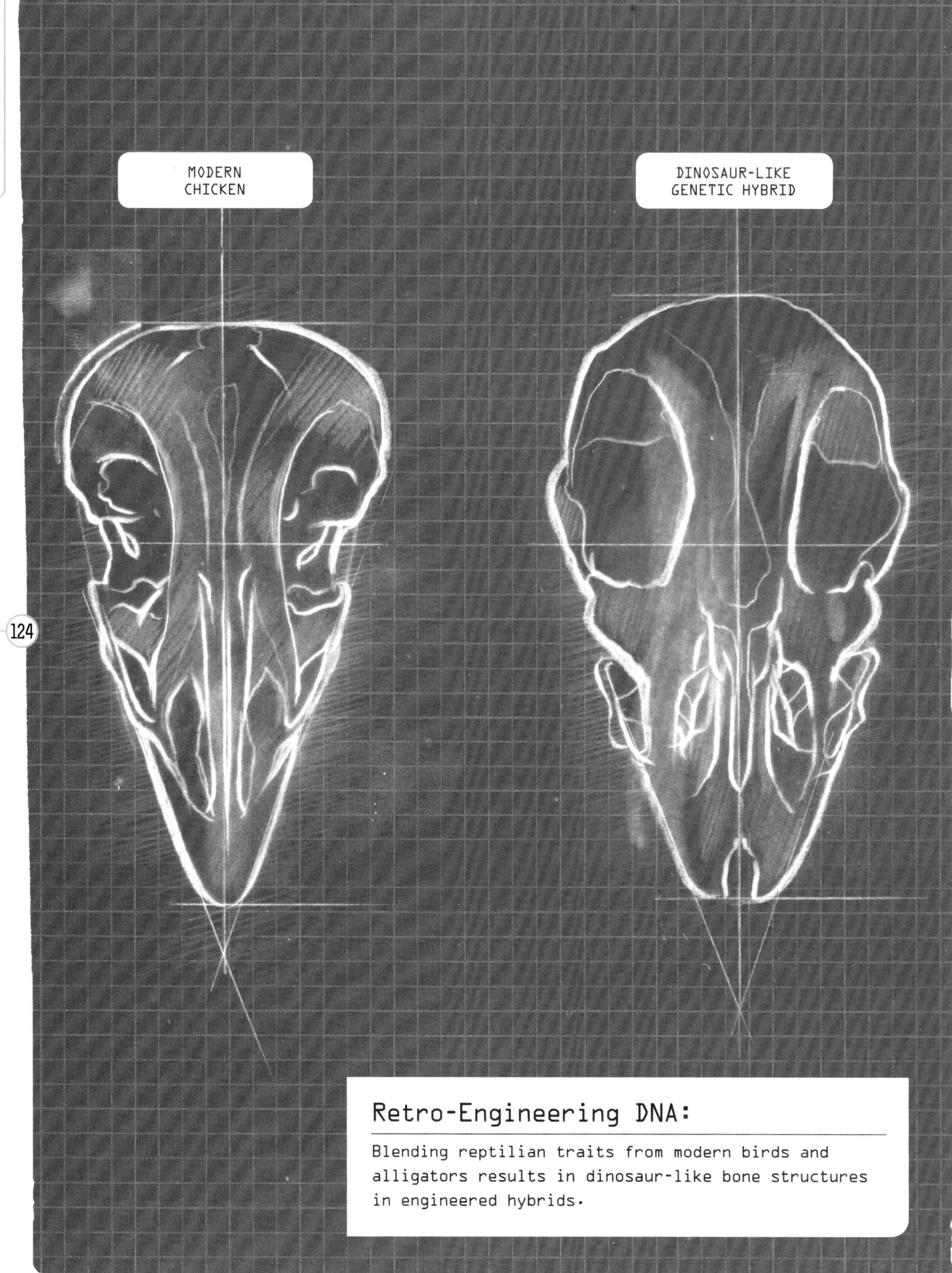

Retro-Engineering DNA:

Blending reptilian traits from modern birds and alligators results in dinosaur-like bone structures in engineered hybrids.

after dropping out of college. Horner has been at the forefront of dinosaur research for decades. His work in the 1970s and 1980s pioneered the idea that dinosaurs could move quickly, work together, and be benign (even nurturing) parents to their young. These were groundbreaking notions at the time and turned the understanding of dinosaurs on its head. What could be more natural, then, for him to attempt to recreate the long-extinct creatures?

Since the publication of *Jurassic Park*, Horner, who has stated his desire to have "a pet *T-rex*" has led the way in efforts to reverse-engineer dinosaurs in real-life, though not in the way that Michael Crichton envisioned. In *Jurassic Park*, preserved dinosaur DNA is altered with current, reptilian, amphibian, and bird DNA to fill gaps in the genetic sequencing caused by the preservation process in the gut of parasitic insects. But in reality, DNA does not preserve well, and extracting the necessary elements from ancient DNA to create a life-sized *T-rex* seems an impossible task. So Horner found another route to recreating the past.

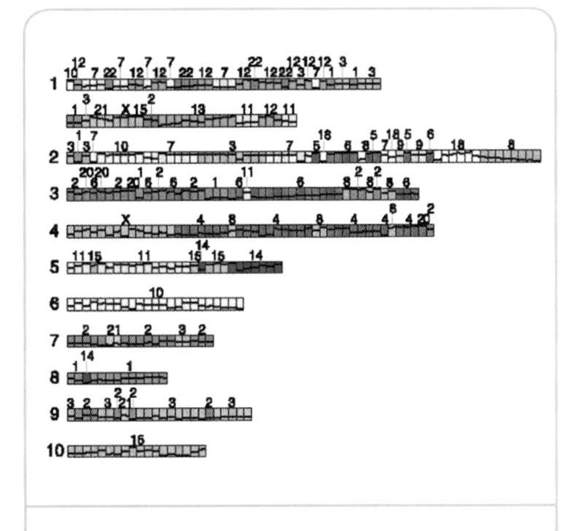

Not One of Us

Mammals and modern birds have very diffferent DNA. Bird DNA (as shown above) is closer to dinosaurs.

'Chickens and all birds are carrying much bigger chunks of dinosaur DNA than we are ever likely to find in the fossil record,' he says. 'DNA is an enormous molecule, made from trillions of pieces, held together in a cell nucleus by chemistry. As soon as the cell dies, that chemistry shuts down, and this molecule, which is very fragile, starts to come apart. It's a process that happens quickly, and we don't think that there would be anything left after millions of years.'

Locked in the DNA of chickens and other modern birds is the same genetic coding that evolved over hundreds of millions of years into the empire of Jurassic-era reptiles. Horner and others have been carefully crafting variants of chicken DNA to replicate the terrible lizards of the distant past. It is a far cry from the *velociraptors* and *Indominus rex* of the movies, but it is still a genetically engineered miracle of innovative thinking.

In *Jurassic Park* and its sequels, dinosaurs are bred and eventually trained for the entertainment of theme park patrons. They are portrayed as intelligent, cunning and in some cases able to work collectively. One of the many questions that remains is: how likely is this scenario is to occur in a real-world breeding program?

'Regarding animal intelligence, we really don't understand it very well,' Horner says. 'We are very mammal-centric...we tend to feel that *our* way of thinking is the best way to do it. Yet we have absolutely no idea how other kinds of animals think or process information. With the *Indominus rex* [of *Jurassic World* fame], we've taken different characteristics from different animals and combined them together. Obviously, if you took some of the processing characteristics from other kinds of animals you would get a better thinker.'

By blending carefully chosen DNA samples, and tinkering with the resulting combinations, scientists can craft reptilian chickens that come closer and closer to resembling dinosaurs. So far, they have managed to turn a beak into a scaly snout and have transformed feathered plumage back into a partial reptilian tail. Horner estimates that they are about halfway to recreating a compact-sized *T-rex*, but notes that like in much of science, advancements tend to come in irregular bursts. It could be three years or thirty before you will see people taking their pet dinosaurs for a walk through the park, seeking unwary pigeons for a

snack. Physically, researchers still need to convert the wings back into arms with hands, lengthen the tail, and add teeth to the snout. But it's just a matter of time.

There is, however, more sleuthing to be done. Although there are dozens of well-preserved dinosaur skeletons in museums and labs around the world, one part of dinosaur creation involves still more fossil-hunting. Scientists have a general understanding of the incremental changes they need to create in the lab, but they need to spot the progressive changes in evolution, in the fossil record of birds. Each change that they seek to engineer in chickens occurred in reverse somewhere along the long evolutionary pathway to modern birds, and the more of these small changes they can find in nature's portfolio, the easier it will be for them to accomplish their task.

But, in ways which echo the unwanted alterations that caused the cinematic versions to wipe out their human hosts, unexpected changes can occur when tampering with genes. A telling example involved the modification of a bird beak to a saurian snout. When the dino-snout experiment was complete, the researchers found that the palate bone inside the mouth had also changed. It broadened and flattened, and connected to the skull in a way that the scientists had not envisioned. Going back to the fossil record, they found that indeed the snout and inner bones had evolved synchronously. It was a small but informative oversight, and fortunately did not result in the lab being destroyed or the devouring of any graduate students. But these are small-scale tests, and nobody knows what might occur in larger genetic modification programs.

Chickenosaurus Reptilianus

By enhancing long-dormant reptilian codes in modern chicken DNA, researchers are preparing to breed tyranno-chickens.

HANDS-FREE: INTERFACES IN AN *IRON MAN* WORLD

Virtual displays and gestural computing are cool, and the virtual computer displays in the movie *Iron Man* were a big hit. And they are very close to reality…or not, depending on what will satisfy you.

LEARNING TO INTERACT

The good news is that you already have a version of this virtual reality on smartphones, tablets, and touchscreen laptops, and it is getting better. The bad news? It's really hard, as it turns out, to make images pop up in thin air. Ironically, if you can do that, it's fairly easy to interact with them.

So we're going to discuss some of that awesome tech in movies such as *Iron Man* and *Minority Report* for manipulating your computer interface with hand gestures—the 'wavy hands' systems. Some use screens, while others are attempts at projections (holographic or other).

The movies make it look so easy and fabulous and studies do actually confirm what the movies suggest—that gesturing is a far more natural and intuitive form of interacting than keyboards, the mouse, touchpads, or touchscreens than our current techniques. And while it's not perfect for every human-computer interaction, gesturing can be more expressive, causes less wear-and-tear on wrists and fingers, and enables the user to focus on the task rather than the input device, making the interaction as natural and transparent as possible.

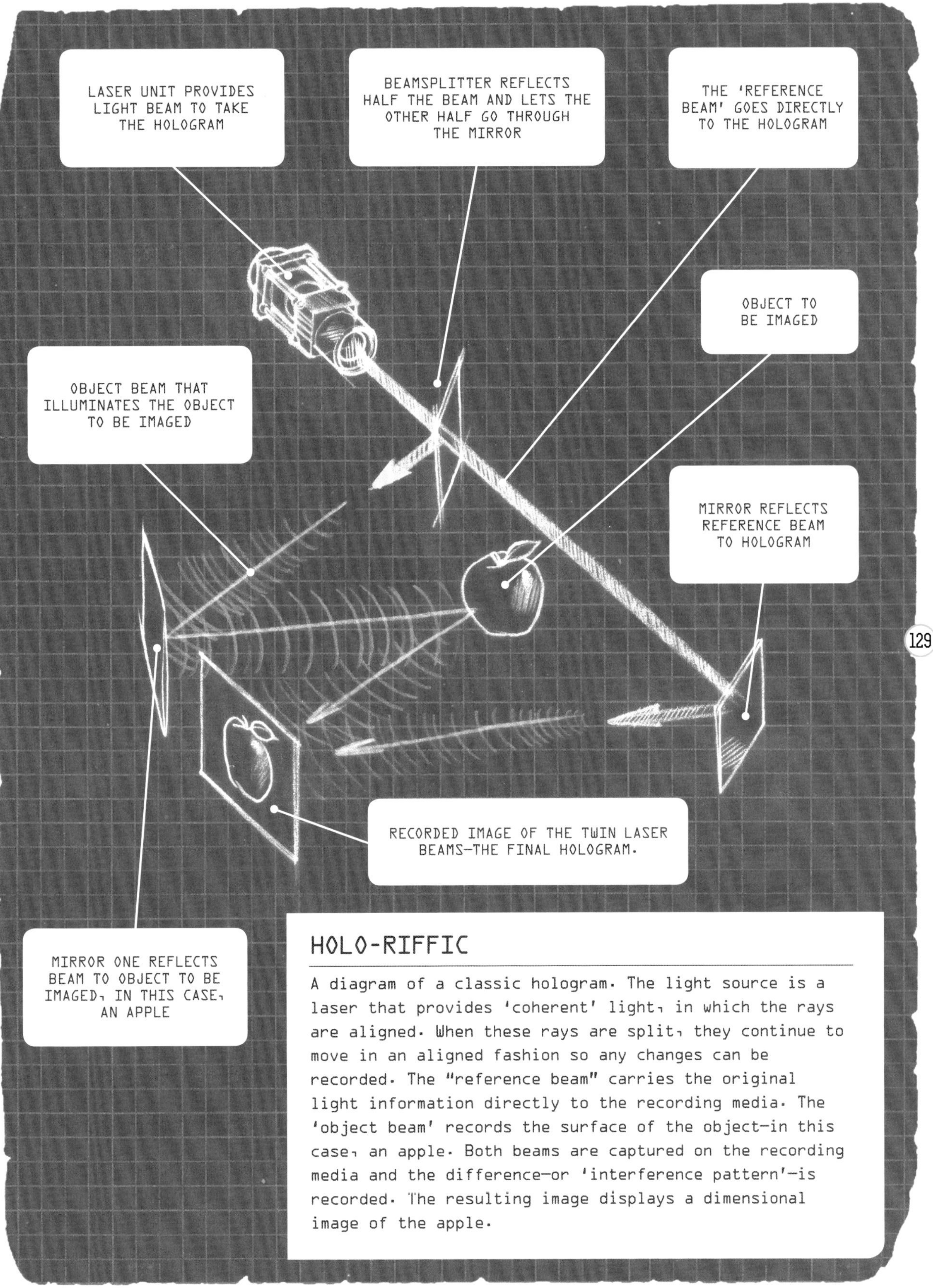

LASER UNIT PROVIDES LIGHT BEAM TO TAKE THE HOLOGRAM

BEAMSPLITTER REFLECTS HALF THE BEAM AND LETS THE OTHER HALF GO THROUGH THE MIRROR

THE 'REFERENCE BEAM' GOES DIRECTLY TO THE HOLOGRAM

OBJECT TO BE IMAGED

OBJECT BEAM THAT ILLUMINATES THE OBJECT TO BE IMAGED

MIRROR REFLECTS REFERENCE BEAM TO HOLOGRAM

RECORDED IMAGE OF THE TWIN LASER BEAMS—THE FINAL HOLOGRAM.

MIRROR ONE REFLECTS BEAM TO OBJECT TO BE IMAGED, IN THIS CASE, AN APPLE

HOLO-RIFFIC

A diagram of a classic hologram. The light source is a laser that provides 'coherent' light, in which the rays are aligned. When these rays are split, they continue to move in an aligned fashion so any changes can be recorded. The "reference beam" carries the original light information directly to the recording media. The 'object beam' records the surface of the object—in this case, an apple. Both beams are captured on the recording media and the difference—or 'interference pattern'—is recorded. The resulting image displays a dimensional image of the apple.

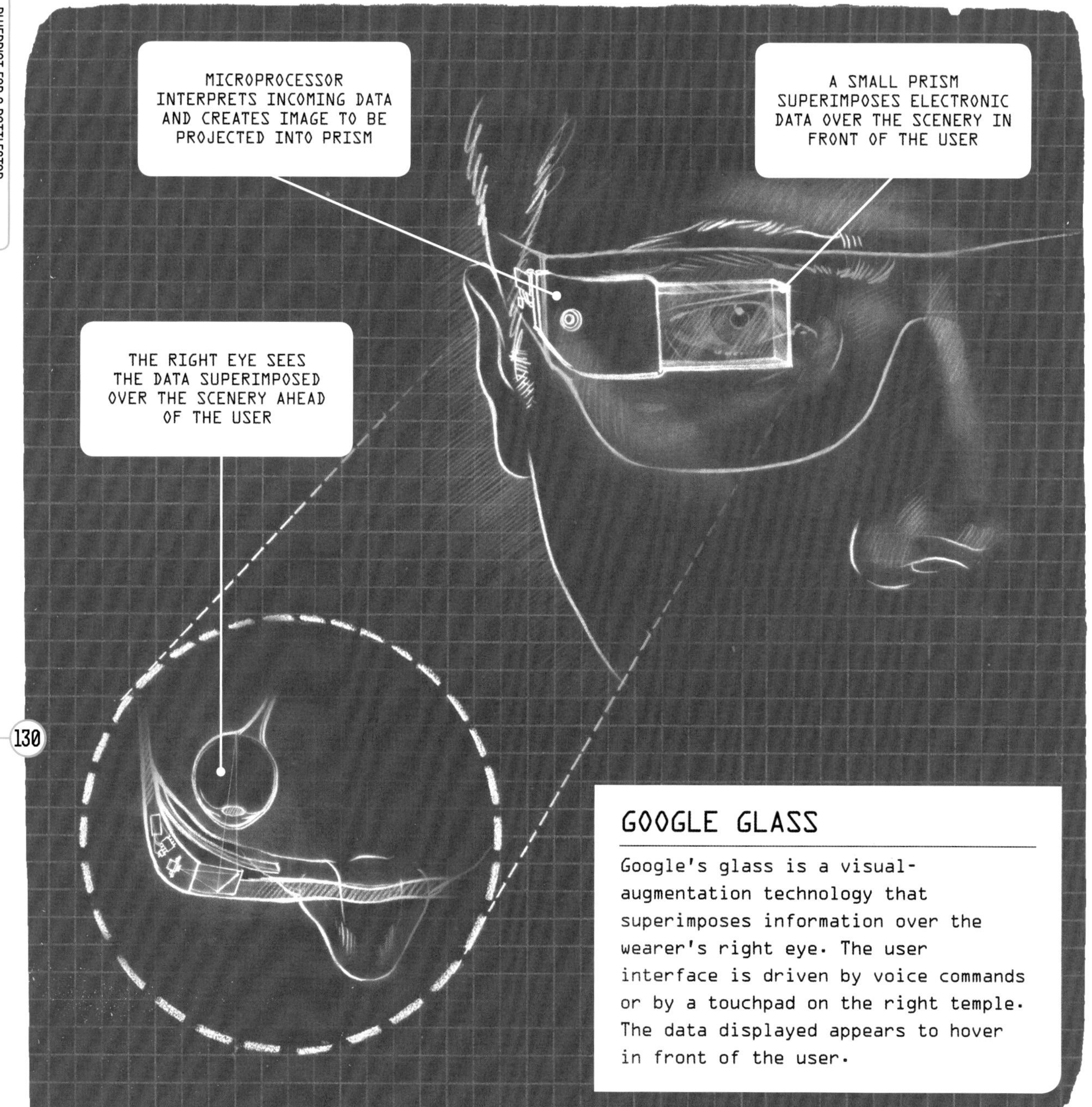

GOOGLE GLASS

Google's glass is a visual-augmentation technology that superimposes information over the wearer's right eye. The user interface is driven by voice commands or by a touchpad on the right temple. The data displayed appears to hover in front of the user.

The visual interface with computers and other interface-driven devices are key to gestural communication. This can already be accomplished with traditional large screens, and the transparent displays seen in high-tech movies and on TV are already available, albeit hugely expensive.

Likewise, there are a few different techniques for tracking hand and body gestures already in use. Some game consoles use controller tracking, but we will discount that for our purposes here because this uses a mechanical device rather than hands to control the computer.

The first widely available motion tracking device was Microsoft's Kinect. First rolled out in 2010, it consists of a small bar, a bit larger than a box of foil wrap, with cameras, depth sensor, and a special microphone facing front. It sits on your TV or game unit *and watches everything you do*—(and by that, I mean that it tracks you to inform the game unit how you are responding to the visual input you are receiving.

Thanks to modern computing power, the Kinect is able to shine an invisible infrared laser into the room, then detect reflection from an object (you) with a light sensor. With this, and proprietary software, the Kinect can perform an analysis of the image it receives, blended with sound interpretation, and distance-ranging information, to create a 3D image map of you. It also performs facial and voice recognition. Newer models have a voice-activated feature (which means that it is *always* listening). It can perform these tasks for up to six people at a time, with up to 20 'joints' or hinge-points, per person (hinge-points refer to places where your body can bend). Oh, and it's motorized so that it can move on its own to better keep track of your movements while playing. That's a phenomenal accomplishment, and only with today's computing power is it possible at all. It's expensive, too: Microsoft reportedly spent almost $500 million on the rollout—or roughly what it cost to launch a space shuttle—to debut the Kinect.

The uses for this technology are not limited to gaming (though that's the consumer golden goose). Gesture tracking is being applied to cars, giving drivers control over the increasingly complex set of options that modern electronic vehicles provide. Such devices are also becoming more commonplace in medical settings, enabling, for example, fully gowned and gloved surgeons to take a break from cutting and sewing to manipulate a video display showing details of the operation in process, or even to carry out the whole operation remotely via robotic manipulators. Kinect-like technology will also find its way into assisting the motion-impaired, tracking facial movements or other parts of the body, to enable a better user experience for people who cannot use more traditional input devices. Gestural tracking is becoming big business.

There are other multiple gesture-tracking and interpretation systems and technologies: some use infrared reflected off the body; others track changes in acoustics as you move; some use multiple cameras and mapping software; and still others utilize electromagnetic fields. Combinations of these approaches may be the best answer of all. The trick is to recognize smaller and smaller gestures to give truly fine-tuned control over the computing device. It is one thing to track an arm motion, quite another to interpret fingers pinching, twisting, and tapping in thin air. Increases in resolution of the pickup devices, together with stronger 3D-mapping and gesture interpretation software, will likely provide this quite soon. For truly minute manipulations, gloves with embedded sensors and transmitters may end up being the most practical solution. All of this will, of course, be combined with voice recognition capability to facilitate an immersive, high-fidelity experience.

Gesture tracking is only half the battle—we still want those big, transparent glass, or holographic displays to which we respond with voice and motion. Transparent video screens are already here and will only get cheaper and better over time, but virtual or holographic image displays have proven to be much more of a challenge and researchers are trying out a wide variety of approaches.

Here are a few of the systems that do not require any special viewing devices, a welcome advance (watching 3D movies with glasses can prove quite a chore).The goal here is for these displays to be realistic, convincing, and unobtrusive. One unique approach to truly 3D displays (also known as 'volumetric' displays) was exhibited by the University of Southern California a few years ago. It uses a mirror about eight inches wide spinning on a turntable, onto which an image is projected. The spinning mirror simulates a fairly convincing 3D space. It sounds simple, but the software that generates the projected image must interpret whatever it is displaying so that is ends up looking like a solid object from various viewing angles—which is a daunting task requiring the creation of up to 5,000 individual images per second. How practical it is for more complex images remains to be seen, but it is a unique way of approaching the problem.

Another approach seems on the face of it even weirder: fog. MistTable, a project at the University of Bristol, uses a table with a hood over it to project an interactive 3D image. The machine creates a liquid mist, or fog, inside the virtual cube created by the cover and tabletop. The 3D image is projected into this. When users reach inside, their hand movements are tracked and they are then able to interact with the imagery in a fairly convincing manner—items can be highlighted, rotated, resized and repositioned. The system needs no goggles or gloves, and multiple people can interact with the virtual image. The gestural interface requires placing your hand within the image, but there is no reason that motion-tracking of the hand could not be made to work from a distance, enabling remote manipulation of objects. The downsides (at least at this stage of development) are that the images are a bit wavy and uncertain—they are being projected into a cloud, after all—and that your hand gets damp. It might prove to be a dead end mechanically, but it's of the high level of innovation in the virtual reality field.

Other approaches include one by Microsoft, which recently patented a device to project a hologram that hovers over the surface of a laptop, tablet or other display with no goggles needed. The holographic system uses cameras to record the surface of the machine being used, then merges detected hand gestures to calculate commands. Face tracking completes the loop, enabling the system to optimize the view for the eyeballs it detects. This also makes a secure-viewing option available—viewers other than the primary user would have to log-in to view the interaction.

The latest 3D imaging systems need a whole new technical vocabulary to describe them. Voxels are basically pixels with volume, aka 3-D pixels. The technology uses cameras, mirrors, and special lasers to create the effect. The lasers are powerful, but they fire in very short bursts (we're talking femtoseconds, a quadrillionth of a second). This creates floating images of up to 200,000 dots per second and these are safe to touch—previous experiments with this kind of system were dangerous to users, as they were hot enough to burn fingers. The images created by this new system utilize glowing plasma that results from the powerful laser 'burning' or ionizing the air. When the image is 'touched' with a fingertip, it is reported to feel a bit like sandpaper, but is so brief that it barely warms the fingertips. The technology is still experimental, but the images created are some of the most vivid and lifelike to date.

A bevy of VR and AR (augmented reality) systems using optical headsets to create a simulated 3D display are in development and on the market, ranging from Google Glass to Oculus Rift and Microsoft's Hololens. These systems are not quite up to *Iron Man* specs, but they are very much the cutting edge for high-fidelity and immersive displays. They can use hand-motion tracking to allow gestural manipulation of virtual objects, but they are not quite the open-air system we are all hoping for. That said, VR and AR worlds are very close to becoming a mass-market product—at this point software is more of a sticking point than the hardware, with Google offering a $15 cardboard head-mounted display that can turn a smartphone into a low-res VR display.

Google Glass is one example of early work in augmented reality (*enhancing* what you see, rather than replacing it with a virtual image). Google Glass is a two-way system—you get visual feedback from your commands, via images projected onto the lens. But its level of intention-gathering is low, and is limited to voice commands and a physical touch sensor. Other augmented reality devices, including Hololens, electronic contact lenses, and even new versions of Google Glass, may yet overcome these limitations.

At the far edge of researchers' dreams, famed futurist Ray Kurtzweil has predicted that by 2030, *truly* immersive VR and possibly gestural computing will be accomplished via sending nanobots into the brain through the blood, linking our neocortex to the cloud. Reality would become almost indistinguishable from virtual reality and humans might spend most of their time experiencing virtual worlds rather than the mundane real one.

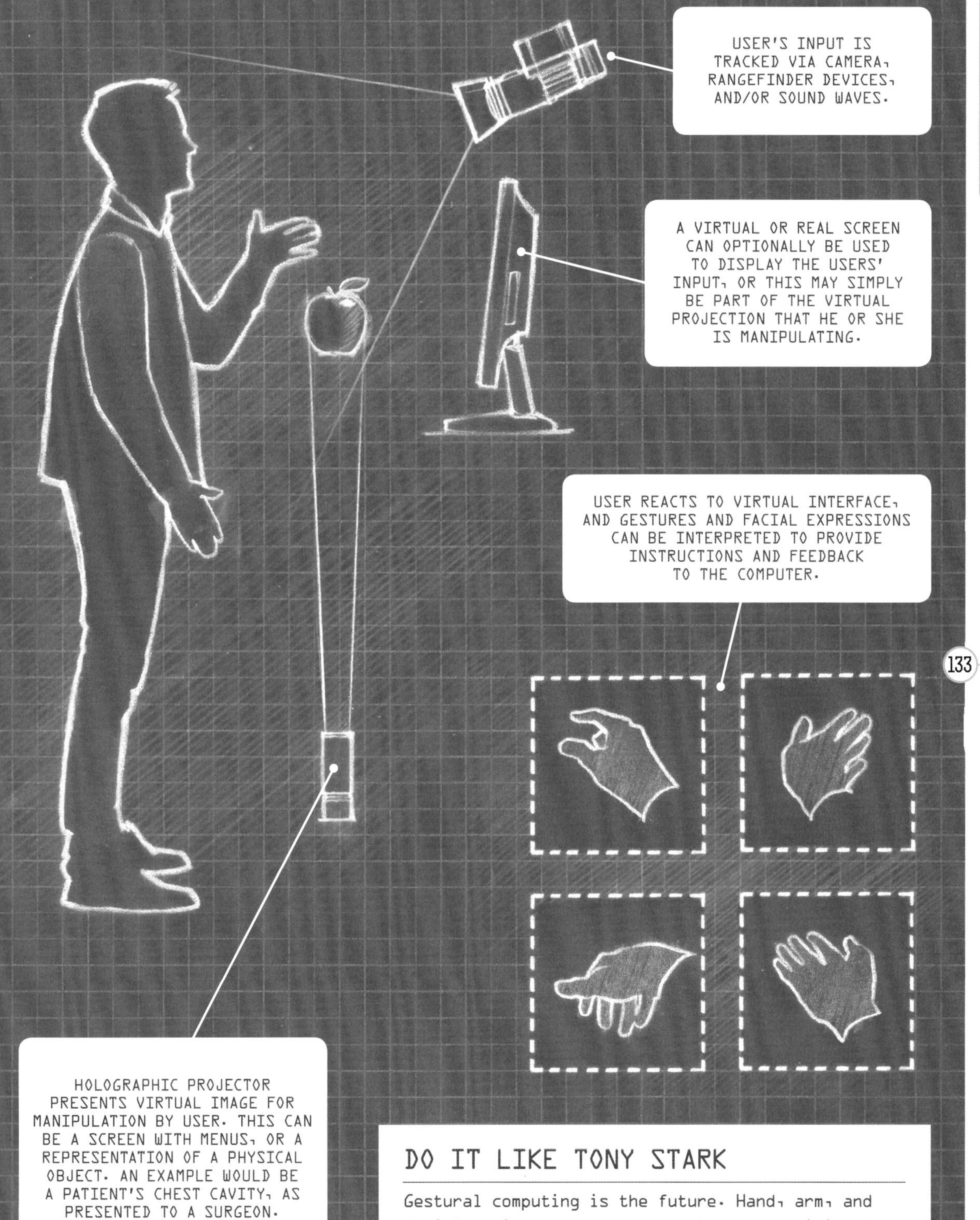

DO IT LIKE TONY STARK

Gestural computing is the future. Hand, arm, and facial motions are more natural than traditional computer input methods and will ultimately become the preferred method of human-computer interaction.

AUTO-DIAGNOSIS: TRICORDERS BEYOND *STAR TREK*

In the original *Star Trek* series, the ship's medic, Dr. McCoy was rarely seen without his tricorder. This handy gadget, a handheld box the size of a large paperback book, provided ready medical analysis for patients—dead or alive. Its very name indicates its usefully wide range of functions— 'tricorder' stands for 'Tri-function Recorder' referring to sensing, computing, and recording.

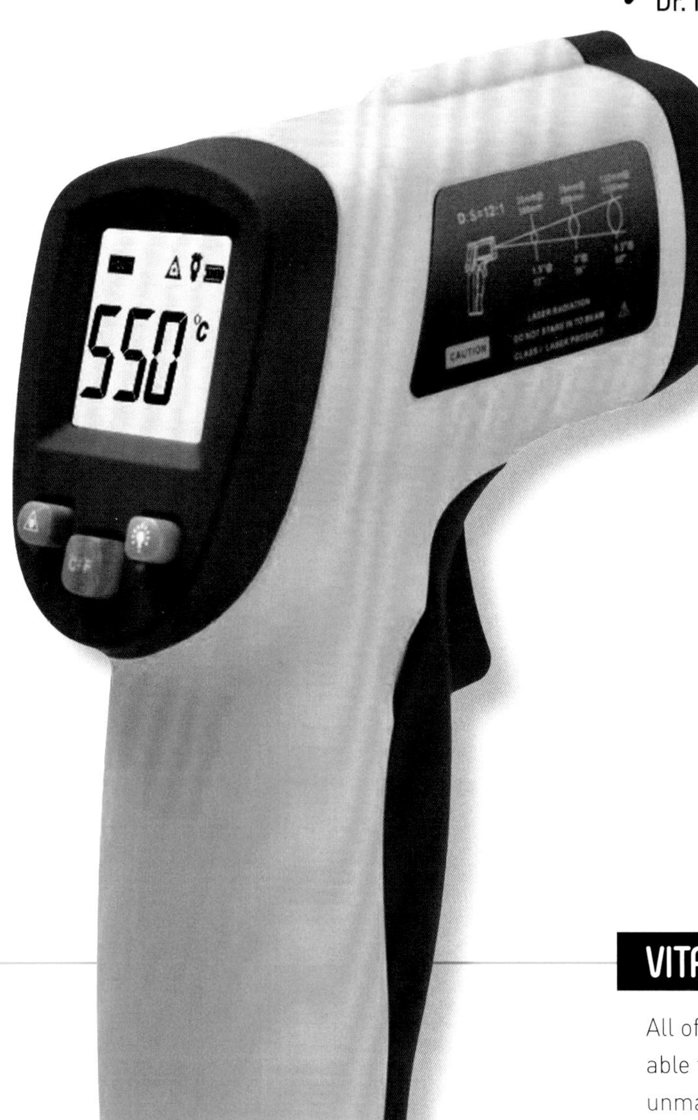

DATA FROM A DISTANCE: MODERN PYROMETERS ARE THERMOMETERS THAT CAN MEASURE TEMPERATURE WITHOUT TOUCHING THE TARGET OBJECT, VIA READING INFRA-RED (THERMAL RADIATION) EMISSIONS.

VITAL SIGNS

All of the countless functions the tricorder was able to perform were controlled by three tiny and unmarked buttons. For 'remote sensing,' the medical tricorder featured a small fingertip-held device housed inside the main unit. This was a body-analysis sensor, a little device that twittered and lit up as it did its work, conveying vital signs to the tricorder. The idea of a portable, handheld diagnostic machine made so much sense that now we are building these things for real.

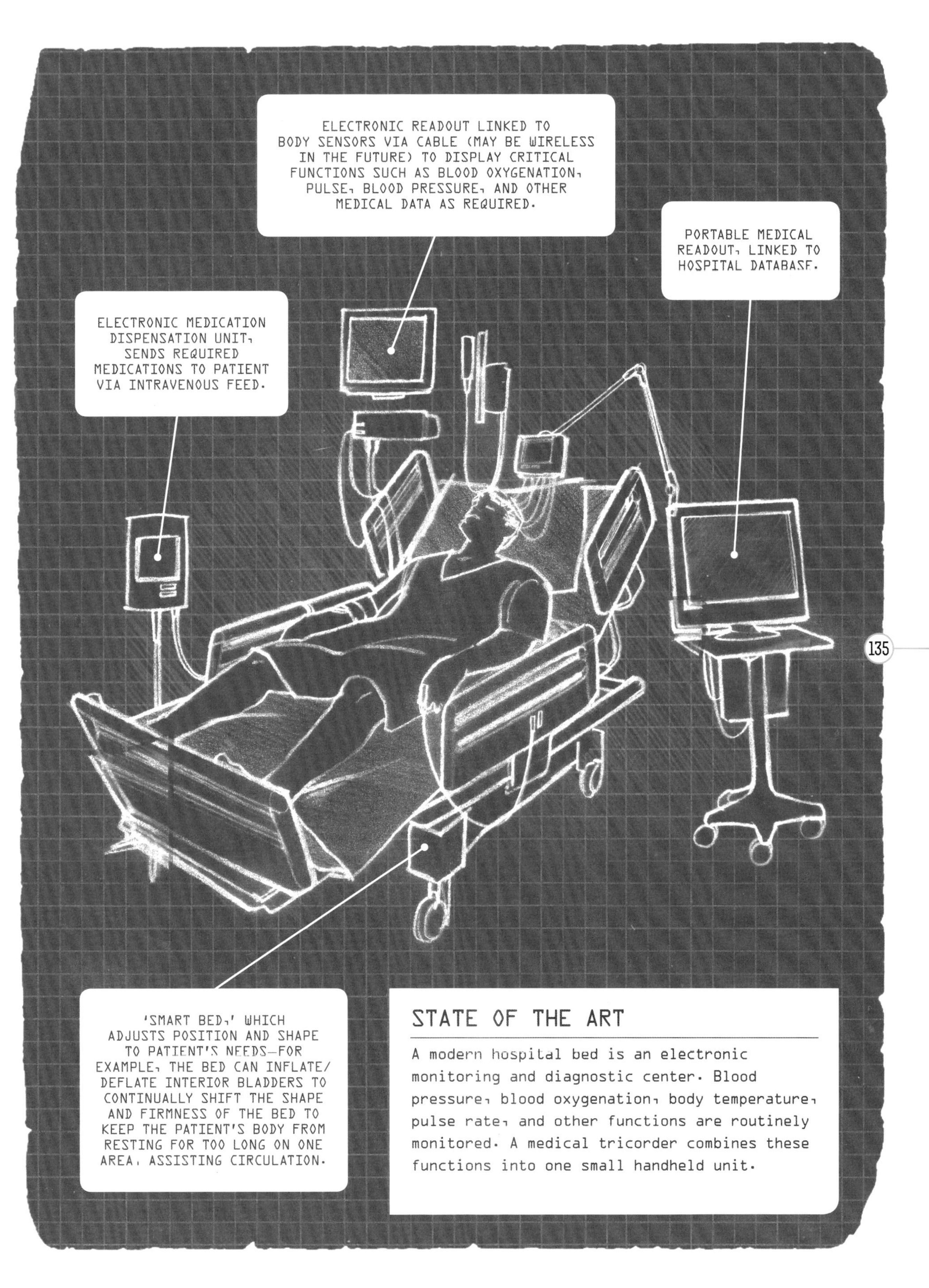

STATE OF THE ART

A modern hospital bed is an electronic monitoring and diagnostic center. Blood pressure, blood oxygenation, body temperature, pulse rate, and other functions are routinely monitored. A medical tricorder combines these functions into one small handheld unit.

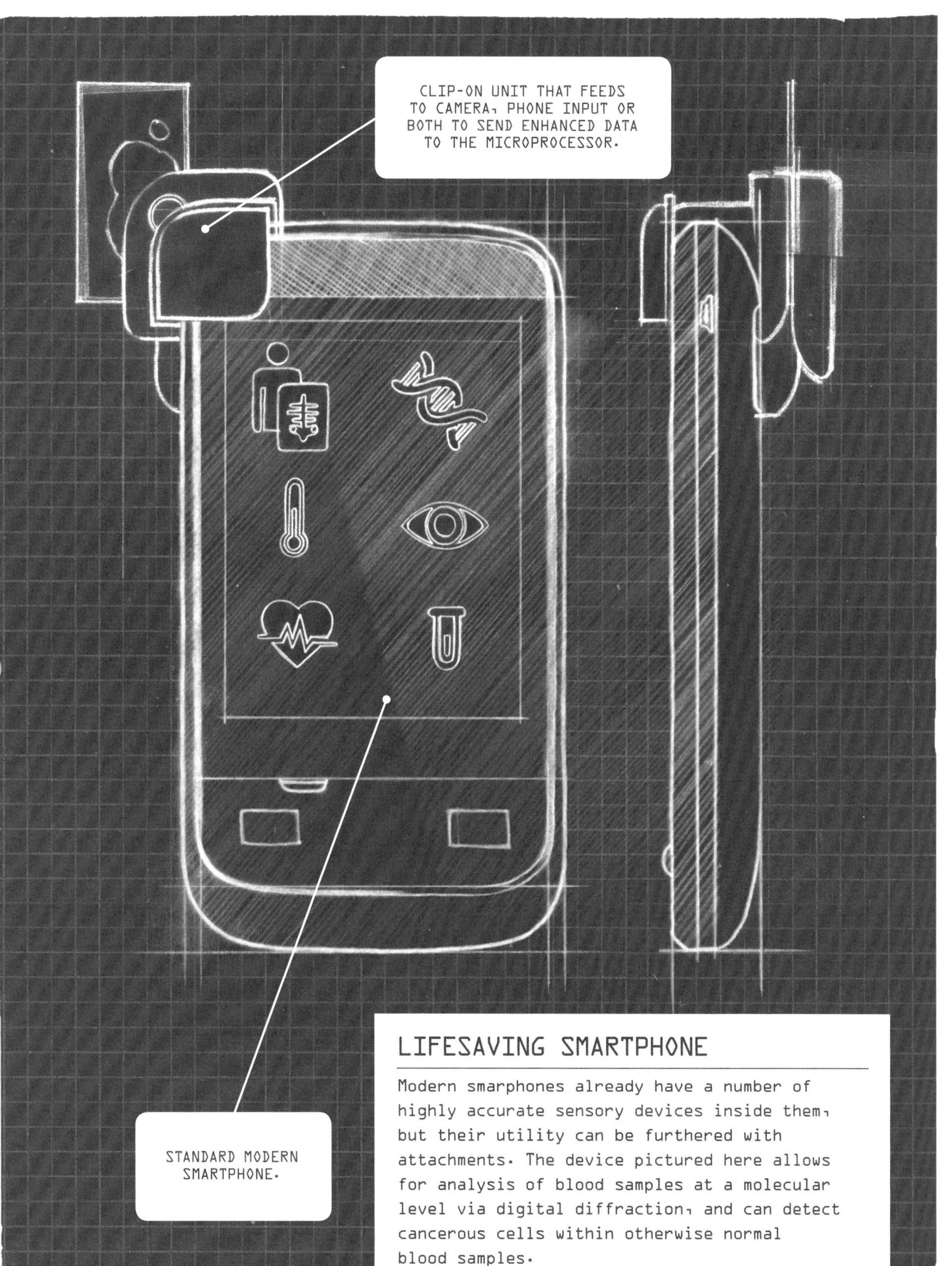

LIFESAVING SMARTPHONE

Modern smarphones already have a number of highly accurate sensory devices inside them, but their utility can be furthered with attachments. The device pictured here allows for analysis of blood samples at a molecular level via digital diffraction, and can detect cancerous cells within otherwise normal blood samples.

This idea is not entirely unique, however. Simple old-style glass thermometers were a type of uni-corder, providing one measurement with no analysis. A newer electrical blood pressure cuff, the kind you strap onto your wrist, could be considered a bi-corder—blood pressure and pulse rate, and with limited computing analysis. So technically, the tricorder is simply a three-function measuring device.

In the 1960s, when most medical measurement tools other than a thermometer or stethoscope were huge, unwieldy, buzzing machines that used vacuum tubes, Dr. McCoy's medical tricorder looked like something miraculous. Today, with the advent of smartphones, it seems a bit less impressive. Each day we carry around more computing power in our pocket or purse than NASA had access to for most of the 1960s. And now these extraordinary minicomputers have been adapted for medical applications, with the birth of the health app.

Today, there are thousands of health and wellness apps. Besides their powerful computers, modern smartphones also incorporate a suite of tiny electronic sensors that allow a wide range of functions, including:

- Proximity—How close is your phone to your face or body?
- Motion—How quickly, and in what direction, is the phone moving?
- Light—How light (or dark) is the environment?
- Compass—In which direction is the phone pointing?
- Gyroscope—What direction in 3-D space is the phone pointing? How long did it take to get from A to B? Combining this sensor with the motion sensor gives smartphones the same inertial guidance capability that NASA's robotic spacecraft use to travel to other planets.

Some smartphones can read fingerprints (and sense your pulse,) measure atmospheric pressure, and even measure temperature (but only of its own interior.) Other sensing devices will be added as the role of the smartphone in our lives continues to evolve.

The addition of external devices can enhance the smartphone's abilities even further. These currently include:

- Sensors for radiation, electromagnetic fields, humidity, and food impurities
- Sensors measuruing air pressure, wind speed, temperature, and other climate variables
- Infrared radiation imagers used as low-light cameras or non-contact thermometers
- Spectrometers, for measuring the elemental makeup of the light reflecting from the measured surface

And, in the medical/fitness area:

- Blood glucose levels meters
- Heart rate and EKG sensors (the latter measures electrical patterns in your heart)
- Blood oxygenation sensors
- Breathalyzers and spirometers (the latter measures lung capacity and can provide early detection of Parkinson's disease)
- Brain scanners
- Ultrasound imaging devices (you can see what your insides look like, too)

There are more coming, but these are the mainstay add-ons available now or in beta testing. Many of the functions of the tricorder are already with us, but these still don't represent a single dedicated piece of handheld hardware that can be used to evaluate a broad swath of specific symptoms or conditions, and possibly save a life on the spot.

There is, though, work being carried out on achieving precisely that. Much of this rapid innovation is being driven by the X-Prize Foundation, which first brought us the Ansari X-Prize created to dispense $10,000,000 to the first non-governmental entity to fly a reusable craft into space, return to Earth and fly again within two weeks. This spawned many other X-Prizes, one of which is the Qualcomm Tricorder X-Prize.

The Qualcomm competition offers $10,000,000 to the first entity to develop a device that can

measure ten key metrics of disease, and has stimulated huge interest from a number of private tech innovators as well as public health bodies and governments. While the first prize is not scheduled to be awarded until 2017, it has already driven the development of a number of devices that meet some of the criteria.

One of these is the Scanadu Scout, a hockey-puck sized device which covers many of the same bases as a fully optioned smartphone, but in one tiny, compact, simple unit. Placing it against the forehead gives a quick readout of heart rate, blood pressure, temperature, blood oxygen level, and an EKG (electrocardiogram). The device's inventor, Walter de Brouwer, was spurred to action when his five-year-old son suffered a head injury. He was shocked to realize that there was no aggregation of brain activity readings or other health data for collective analysis, which could help track his son's recovery progress. So began his quest to develop a device that can take quick and easy measurements and aggregate them to a large database for analysis, providing insight on disease prevention and treatment.

So how does the Scanadu work? Measuring blood pressure has always required some kind of cuff around the arm...temperature requires a thermal probe...and an EKG readout has traditionally involved a dozen or so sticky electrical sensors taped all over your chest. Through the ingenious utilization of low-voltage sensing and the power of its microprocessor, the Scanadu completes these previously invasive tasks with a minimum of fuss. Measuring your temperature is pretty straightforward: when you touch the machine to your forehead, the thermal sensor gets all the data it needs to measure that (at least local to the forehead.)

Now, because you are holding the unit to your head, with one fingertip on a sensor placed at the top of the unit (where the forefinger sits) and another on the bottom (where the thumb grips it,) there is a completed electrical circuit—hand→Scanadu→head. This allows the EKG to be created (the heart is primarily a little electrically fired pump, after all). Then, on one of the fingertip grips on the case, there is another sensor called a photoplethysmograph (PPG,) which can read blood flow and oxygenation.

Another product, the AcuPebble, is about the size of a quarter and tracks sleep apnea, chronic obstructive pulmonary disease, asthma, and atrial fibrillation (heartbeat irregularity) as the user sleeps. In conjunction with a smartphone, it can record and track this data and even send it to medical practitioners. The AcuPebble works by listening to your breathing—it mounts on the chest with an adhesive dot and is able to gather the data it needs by simply eavesdropping on your respiration as you slumber. While even a step further away from the universal tricorder in form, the function—to detect and analyze medical conditions—is the same.

The SensoDX takes things a step further than the Scanadu or Pebble, mounting an entire diagnostic lab onto a single chip. There are actually two units: the handheld collection device and a larger analysis unit. It looks more like a mini-ATM than a tricorder, but the machine is able to detect biomarkers, and therefore diagnose such disorders as heart disease, cancer, or drug overdose. The functions of the SensoDX are far more complex than previous products, requiring blood or saliva samples loaded into cartridges and then inserted into the larger analytical machine. Its size, though, renders it more like the medical sensor beds in *Star Trek* than a tricorder.

These devices represent a revolution in medicine, allowing far more preventative and diagnostic work to be done by fewer, far less expensive practitioners. Even simpler versions, such as the Fitbit wristband, are something like a medical tricorder. How much you walk (or don't), how many stairs you climbed (or didn't), how well you slept (or failed to), are all reported. And because the Fitbit conveys this information via Bluetooth, everything is tracked and analyzed in your phone, providing you with information about walking patterns, your calorific burn rate, and how this coordinates with your sleep cycle.

There was another type of tricorder in *Star Trek*, one that registered general scientific as opposed to purely medical readings. There has been one attempt to create such a scientific tricorder. It was undertaken by Peter Jansen, then a student at McMaster University in Ontario, Canada and now a

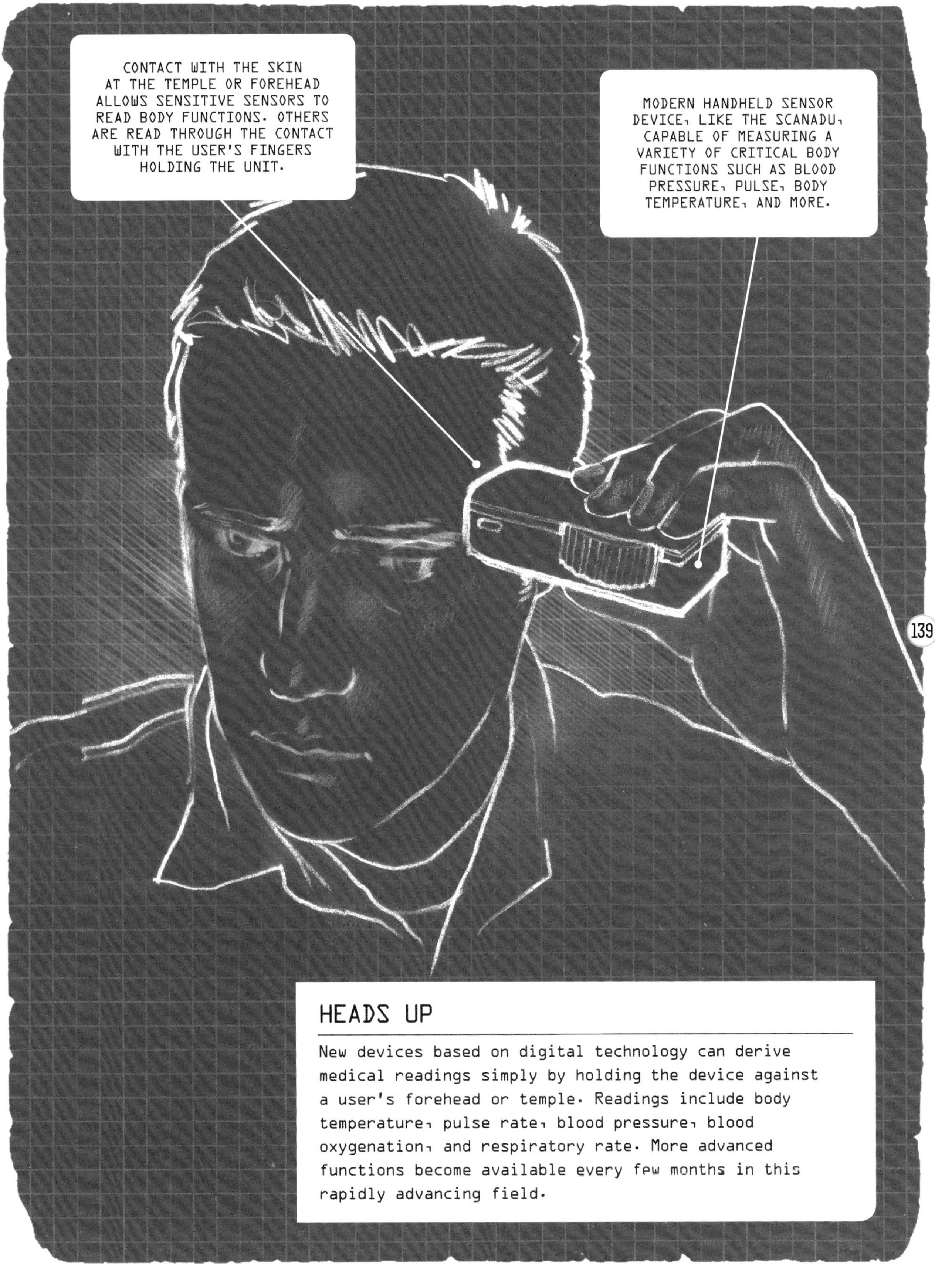

CONTACT WITH THE SKIN AT THE TEMPLE OR FOREHEAD ALLOWS SENSITIVE SENSORS TO READ BODY FUNCTIONS. OTHERS ARE READ THROUGH THE CONTACT WITH THE USER'S FINGERS HOLDING THE UNIT.

MODERN HANDHELD SENSOR DEVICE, LIKE THE SCANADU, CAPABLE OF MEASURING A VARIETY OF CRITICAL BODY FUNCTIONS SUCH AS BLOOD PRESSURE, PULSE, BODY TEMPERATURE, AND MORE.

HEADS UP

New devices based on digital technology can derive medical readings simply by holding the device against a user's forehead or temple. Readings include body temperature, pulse rate, blood pressure, blood oxygenation, and respiratory rate. More advanced functions become available every few months in this rapidly advancing field.

researcher at the University of Arizona, with interests ranging from general cognitive sciences to —more specific to us—sensors. He began his quest way back in 2007 with some ground rules—the device would have to be both affordable and accessible.

As Jansen explains:

The Tricorder project emphasizes accessibility. The devices we build are meant to be as inexpensive as possible, so folks might have access to them without having to worry about the cost, or their difficulty of use. My hope is that someday every household—and every child who wants one—might have access to a small device that can easily be kept close in a pocket or bag, and quickly pulled out when curiosity strikes. By turning a walk home through the park into a nature walk, and dad's springtime home repairs into a lesson about heat flow, it's my hope that everyday experiences will become opportunities to learn and develop an intuitive understanding and deep fluency with the science of our everyday world.

Jansen's earliest efforts were at a time when smartphones were still working pretty hard to simply be good telephones. His Science Tricorder Mk 1 was able to sense the following:

- Atmospheric measurements: temperature, pressure, and humidity
- Electromagnetic measurements: magnetic fields, infrared (temperature,) light colorimeter (degrees Kelvin,) and light level
- Spatial: ultrasound (distance,) GPS, and an inertial measurement unit

Jansen's tricorder even looked very much like one from *Star Trek: The Next Generation*. He also developed his own graphic representations of the various measurements and spent hours measuring electrical leakage from house wiring and wall outlets and even the humidity in his breath.

Jansen continues to work on his tricorder designs, all of which are open-source and thus available to anyone who wants to come and play. He is striving to come up with a one-stop unit that would include the ability to detect the chemical and elemental composition of a target. Peter recently got a delightful surprise when one of his new, experimental models detected unexpected radiation, and he learned from a colleague that there had been a solar flare that morning—his tricorder had spotted it.

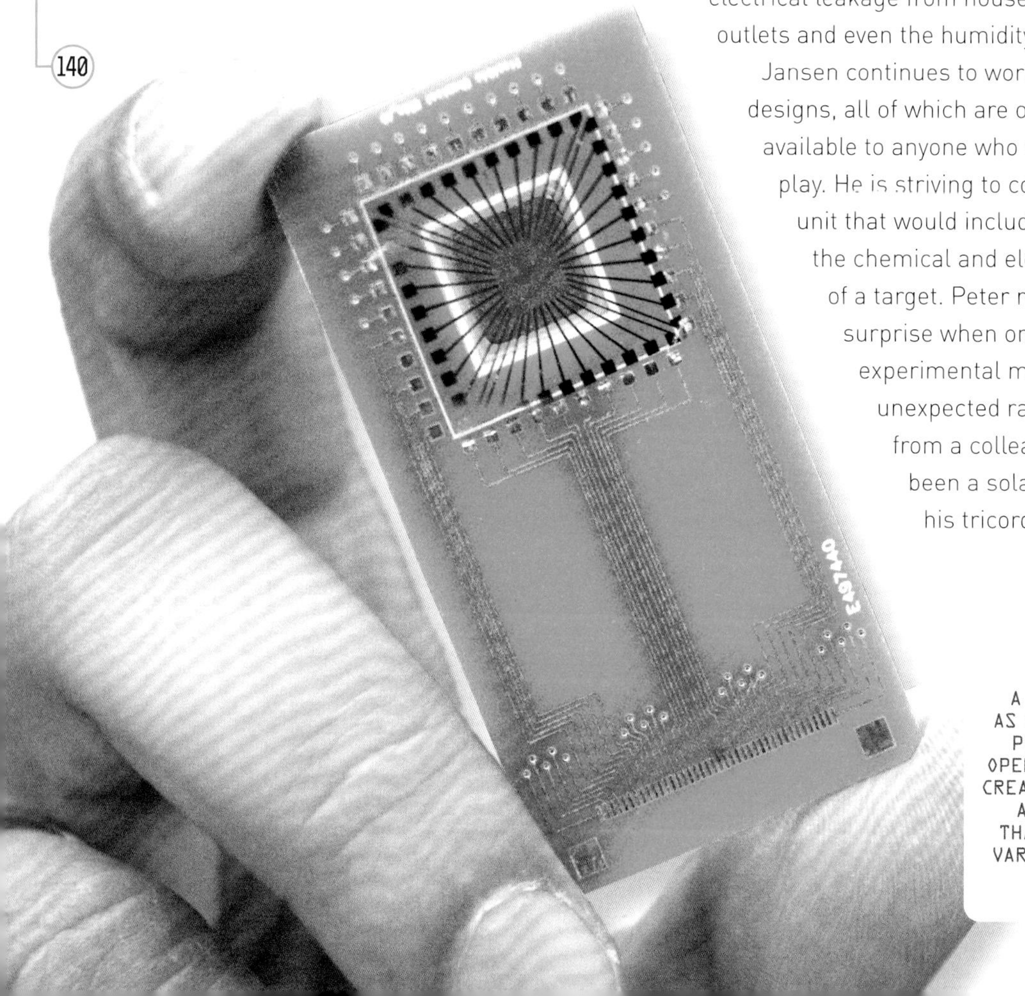

A MICROPROCESSOR CHIP AS USED IN THE TRICORDER PROJECT, A COMPLETELY OPEN-SOURCE INITIATIVE TO CREATE A TRULY UNIVERSALLY ACCESSIBLE TRICORDER THAT CAN FULFILL A WIDE VARIETY OF INVESTIGATIVE FUNCTIONS.

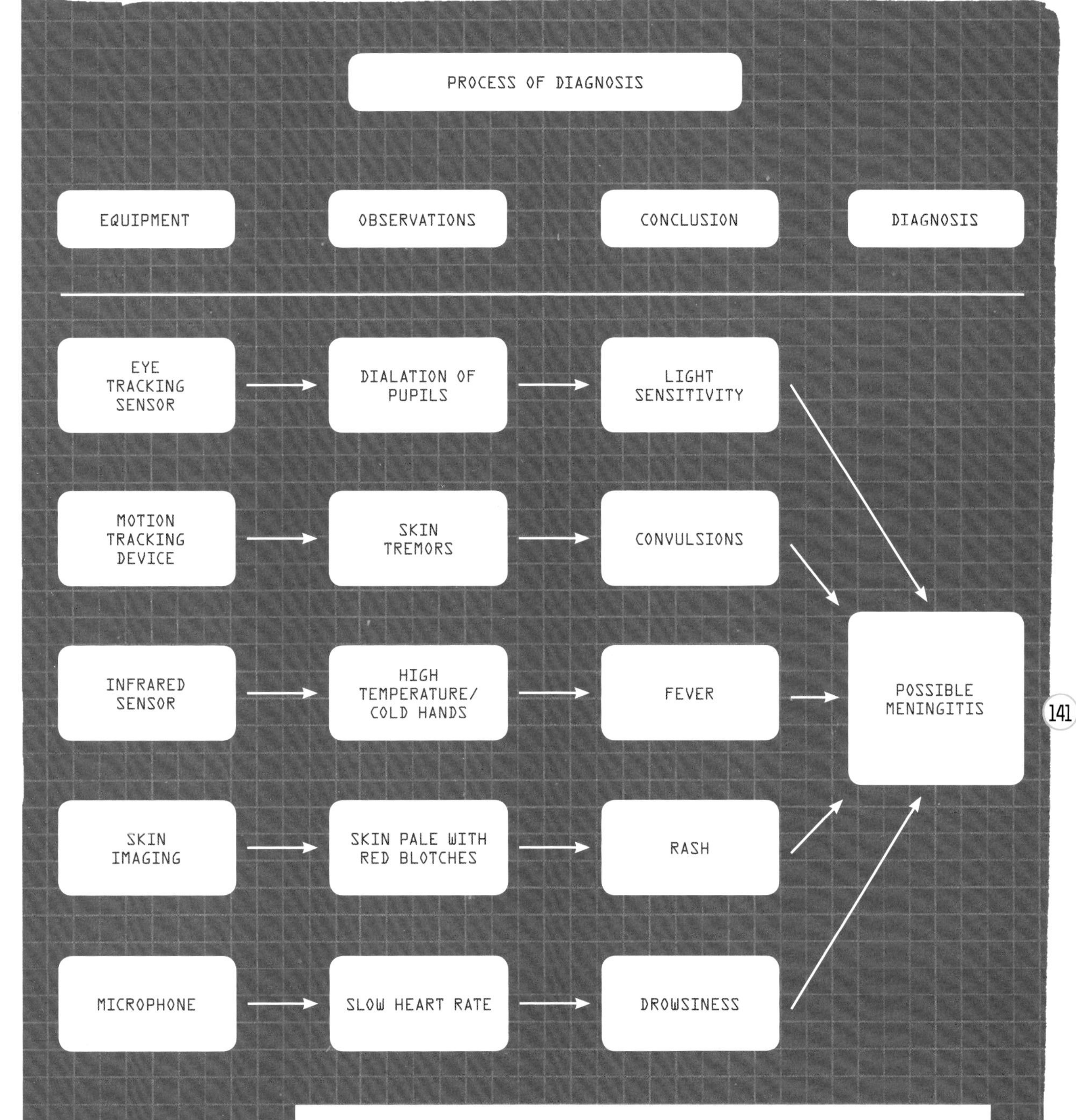

WHAT'S WRONG?

Modern medicine is all about diagnosis—there are so many specific therapies available, that collating and understanding symptoms is critical. Medical tricorders gather an array of measurements (center), supplied by instrumentation (to left), to come up with a diagnosis (at right). The conclusion in this example, possible meningitis, would suggest immediate attention at a medical facility for diagnosis and treatment.

ECO-ENGINEERING: THE SCIENCE OF TERRAFORMING

Mars, our nearest planetary neighbor, is a hellish place: a cold, dry desert word. Living there would be a harsh and relentless struggle in comparison to Earth, which is still in reasonably good shape. But what if we could make Mars, or even another Earth-like planet orbiting a distant star, sufficiently earthlike to become an Earth Mark II?

HUMANITY'S NEW HOME?

That's terraforming in a nutshell—the process of artificially altering another world to make it more suited to human habitation. It is a subject that has been studied a lot, and we now have a basic understanding of how we might go about achieving it. Dramatically altering the climate and conditions of an entire planet will not be cheap or simple, but the recent acceleration of climate change on Earth has given some hints about how these changes occur. In a way our planet is experiencing 'reverse terraforming'—making Earth more Venus-like—although we have a long way to go before our home world enjoys an average daytime temperature of 900°F, an atmospheric pressure almost 100 times that of sea level on Earth, and carbon dioxide-dominated air.

There are people who don't think that terraforming is such a good idea. Some believe that traveling to yet another planet and altering it to suit our needs is the ultimate form of human hubris. Others argue that we would have to be very certain that there are no resident creatures there—to some, even microbes count—that would be erased by our efforts.

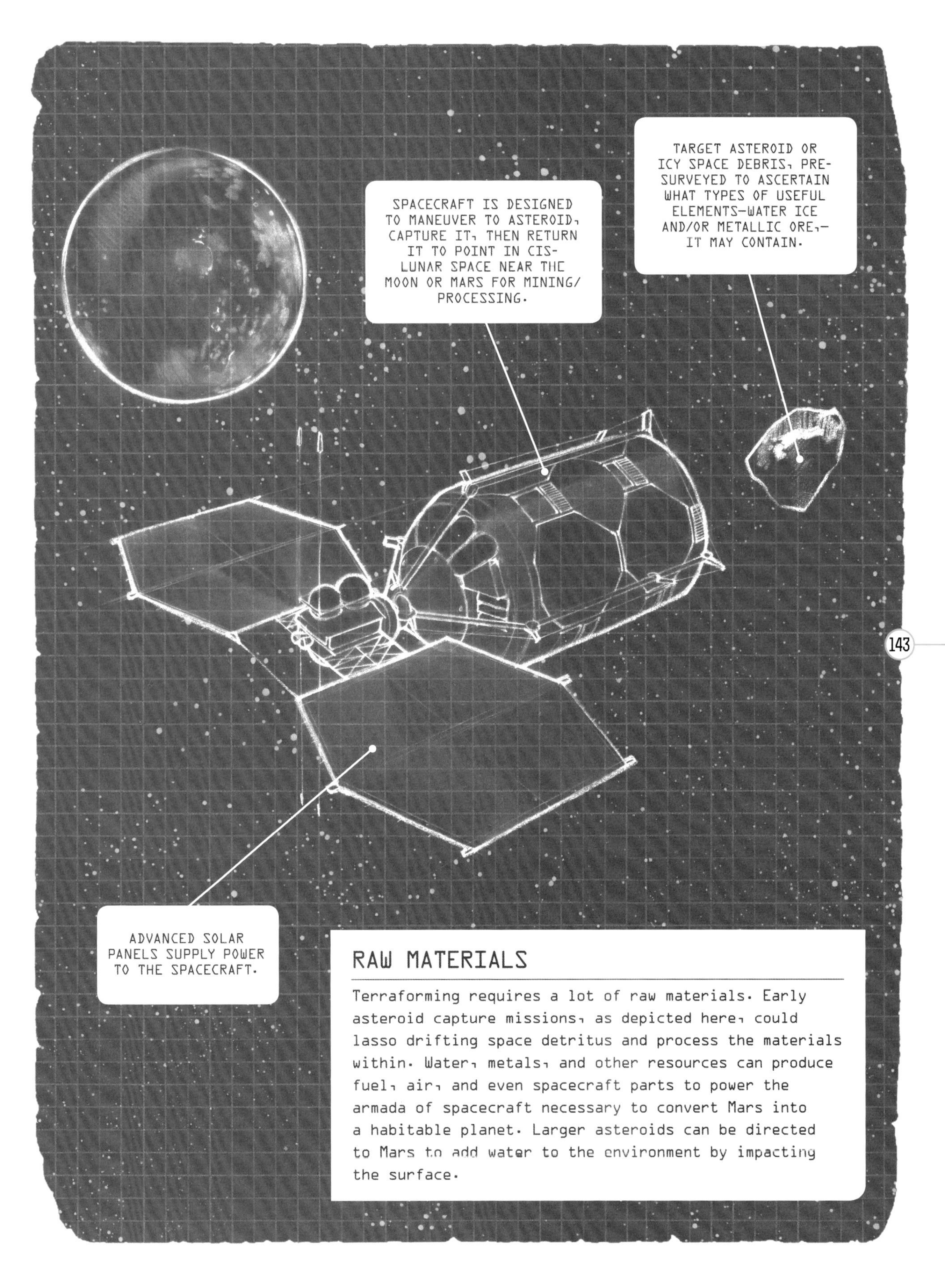

RAW MATERIALS

Terraforming requires a lot of raw materials. Early asteroid capture missions, as depicted here, could lasso drifting space detritus and process the materials within. Water, metals, and other resources can produce fuel, air, and even spacecraft parts to power the armada of spacecraft necessary to convert Mars into a habitable planet. Larger asteroids can be directed to Mars to add water to the environment by impacting the surface.

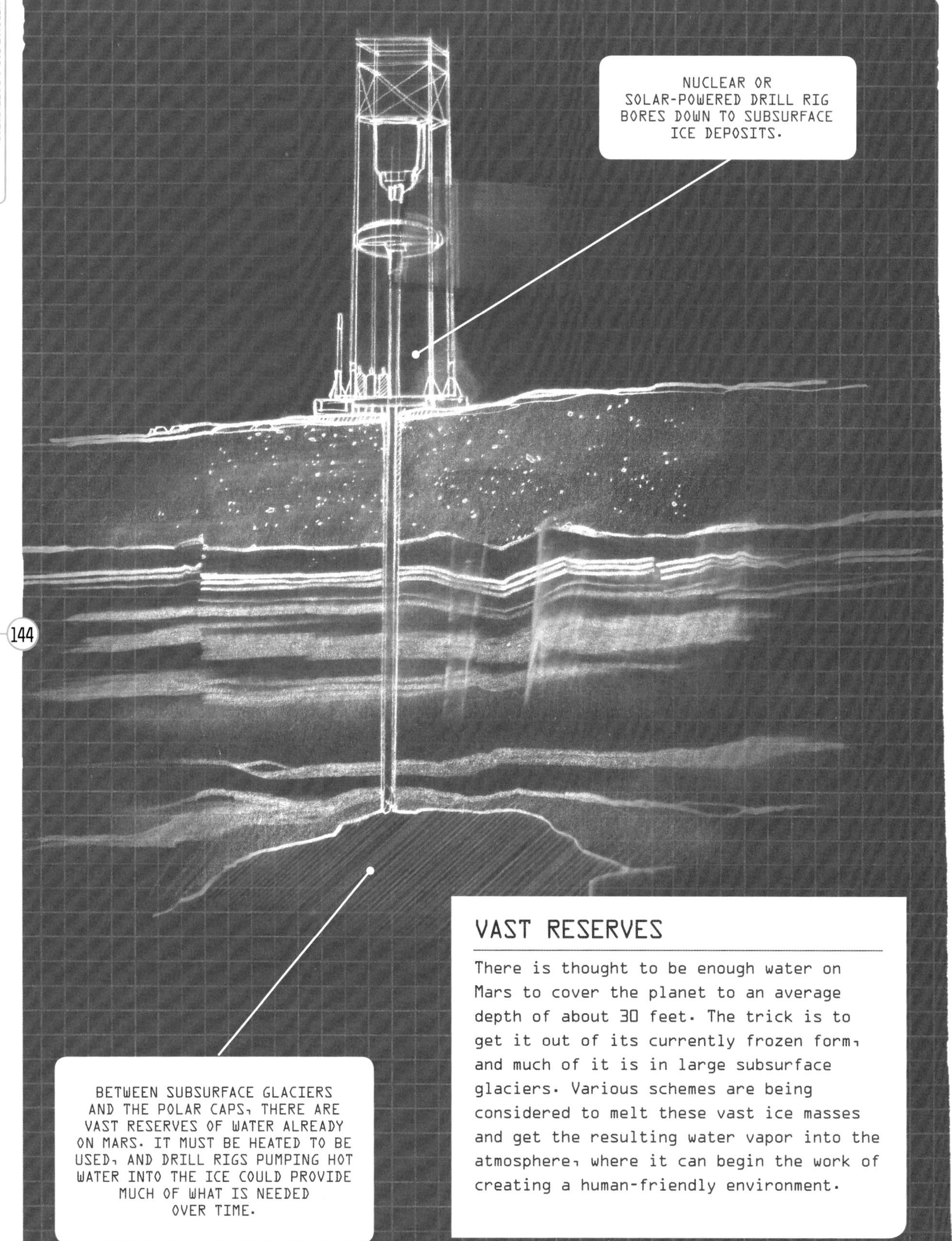

VAST RESERVES

There is thought to be enough water on Mars to cover the planet to an average depth of about 30 feet. The trick is to get it out of its currently frozen form, and much of it is in large subsurface glaciers. Various schemes are being considered to melt these vast ice masses and get the resulting water vapor into the atmosphere, where it can begin the work of creating a human-friendly environment.

But a far larger faction is less concerned, and for good reasons. The pro-terraformers point out that we are depleting our own planet of natural resources so quickly that an alternative is urgently needed and (perhaps optimistically) that we are more enlightened about settlement than we were a century ago (when much of the American West was 'Euroformed').

The pro-terraformers have another excellent point: humanity may not be able to survive on Earth indefinitely. One need look no further than Tunguska, Siberia or Meteor Crater, Arizona to see that there can be surprises from space. Such an asteroid impact may have caused the dinosaurs to become extinct, and another might do the same to humanity.

Assuming that we are serious about creating a Plan B, a new home, or a nearby habitable 'ark,' Mars makes the best choice. The only other relatively local alternative is Venus, but that planet has experienced a runaway greenhouse effect, which has created a hellish environment there. Living on Venus could likely be accomplished only via city-colonies floating atop the densest part of the atmosphere, which doesn't count as terraforming.

Terraforming Mars would, in one sense, return it to what it once was. Since the Curiosity Rover landed on Mars in 2012, it has become clear that the planet was significantly more Earth-like about 3.5 billion years ago, when there were copious amounts of water on its surface in rivers, lakes, and seas. The atmosphere was much denser, and was apparently oxygen-rich. But much of this atmosphere escaped, and with it went the bodies of liquid water. Mars eventually became the cold, dry desert world we see today.

Scientists and engineers at NASA and elsewhere have thought hard about how to go about altering Mars akin to its former state. The key is to create or free up large amounts of greenhouse gases that can boost the atmospheric density to warm the planet and trap moisture. While carbon dioxide is a natural choice, methane or ammonia would work faster. Methane has been found to occur on Mars in small quantities, but its source is as yet unclear. It could be produced by microbial life, or be due to geological processes. If the cause can be identified and the process expanded, this would help towards our terraforming goals. Once the planet is warm enough, vast reserves of frozen water would begin to melt from permafrost, glaciers, and the polar caps. Modern estimates suggest that there is enough water locked-up in ice to cover perhaps 20 percent of the surface in seas and oceans. Freeing some of this would release more carbon dioxide, producing a self-reinforcing runaway grenhouse effect. Thus Mars itself helps the terraforming cause, making less work for the terraformers.

Other schemes include bombarding the planet with icy bodies already found in space, or detonating nuclear explosives to melt the frozen water, possibly by flash-evaporating parts of the polar caps. SpaceX entrepreneur Elon Musk—a proponent of Mars settlement—proposes setting off a rapid-fire series of nuclear bombs over the poles to produce carbon dioxide. These detonations would occur high in the atmosphere to melt the ice below, like bringing a series of small suns to the Martian neighborhood. This in turn should kickstart a greenhouse effect. Even then, once the carbon dioxide has been liberated to do its job of warming the planet, a method must be found to create enough oxygen to make it hospitable to human life. Since this may well take a lot of time—possibly on the order of many tens of thousands of years, it may be easier to adapt humans to breathe a different atmosphere, one richer in carbon dioxide (probably augmented by portable oxygen packs).

Mars has a far lower gravity than Earth, about 38 percent, so the atmospheric mass would have to be proportionally larger than our own to be sufficiently dense to retain liquid water and ultimately become breathable. Also, broad estimates of the mass of usable carbon dioxide contained in the south pole and global surface ice mass of Mars would be only enough to get us halfway to our atmospheric pressure goal of 14 PSI (Earth's sea level). So an external source of gases would be needed to assist in the completion of our task.

One way of achieving this might be by bringing ice to Mars, or finding other ways to liberate similar gasses via the processing of Martian rocks and soil.. The solar system is filled with space junk, mostly in the form of ice and rocks leftover from the formation of the planets. The asteroid belt lies just beyond Mars, and there are huge amounts of icy junk there. If a way could be found to redirect some of that mass, or some of the comets (also largely composed of frozen ice) that routinely patrol the inner solar system, you could send these watery masses smashing to Mars and get the process started that way.

The spectacle of a ten-mile wide cometary nucleus crashing into Mars would likely not sit well with interplanetary environmental activists. Slowly altering the planet's atmosphere is one thing; exploding nukes or smashing comets and asteroids into it is something else. One alternative would be to place an icy body in a looping, grazing orbit of Mars—we already do this to slow down our Mars probes via the friction encountered by skimming the atmosphere. The object would continue to dip in and out of the Martian atmosphere, with some of it being melted away in each pass. Eventually it would release most of its bulk as melted vapor.

Once the atmosphere has thickened with carbon dioxide, some hearty plants will be very much at home. Lichens, fungus, and algae have been suggested as good stock to get the jump-start the Martian ecosystem. This leaves us with another thorny issue: radiation. A denser atmosphere alone will not mitigate all the harmful energy coming to Mars from the sun. On Earth, we are protected by our magnetic field and and atmosphere. Mars has neither. An atmosphere is something that can be engineered; a magnetic field would be far tougher to accomplish. In the end, we will probably need to stay in radiation shelters most of the time to avoid damage to our bodies. But that's not a huge problem—many Mars habitat designs already include the idea of burrowing underground (or building habitats in caves and lava tubes) to provide both radiation protection and sturdy habitats.

Terraforming may well be the lead to the future resettlement of a portion of humanity—top minds such as Stephen Hawking have decreed it a necessity. Hawking has opined that humanity won't survive another thousand years without 'escaping from our fragile planet.' And this may be true. We face many threats to our survival such as resource depletion, a massive asteroid strike on Earth, or an unforeseen eco-catastrophe. These are all scary prospects, but the worst challenges involve human-created threats. If we cannot solve these, migrating to Mars may simply result in our taking the same old problems along with us.

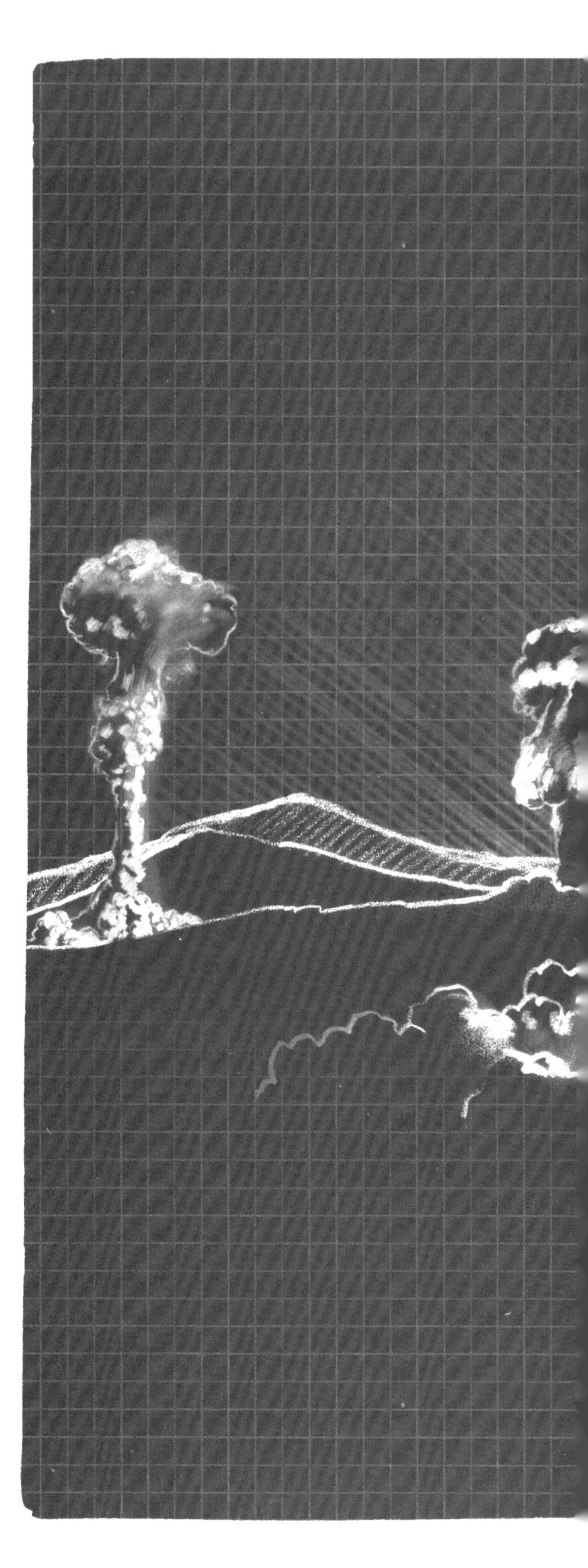

SOME ADVOCATES OF TERRAFORMING MARS, NOTABLY ELON MUSK, HAVE SUGGESTED THE USE OF THERMONUCLEAR EXPLOSIVES NEAR THE POLES TO LIBERATE GASSES BY VAPORIZING ICE MASSES, TO AID THE PROCESS. OTHER METHODS, SUCH AS REDIRECTING SPACE DEBRIS, ARE THOUGHT TO BE POTENTIALLY MORE PRODUCTIVE.

MOBILITY WILL BE IMPORTANT, AND CREWED MARS VEHICLES WILL TAKE VARIOUS FORMS—SOME INCLUDING SUPPLIES, LIFE SUPPORT, AND RADIATION SHIELDING FOR LONG-TERM TRAVEL.
SUBTERRANEAN RADIATION-HARDENED STRUCTURE.

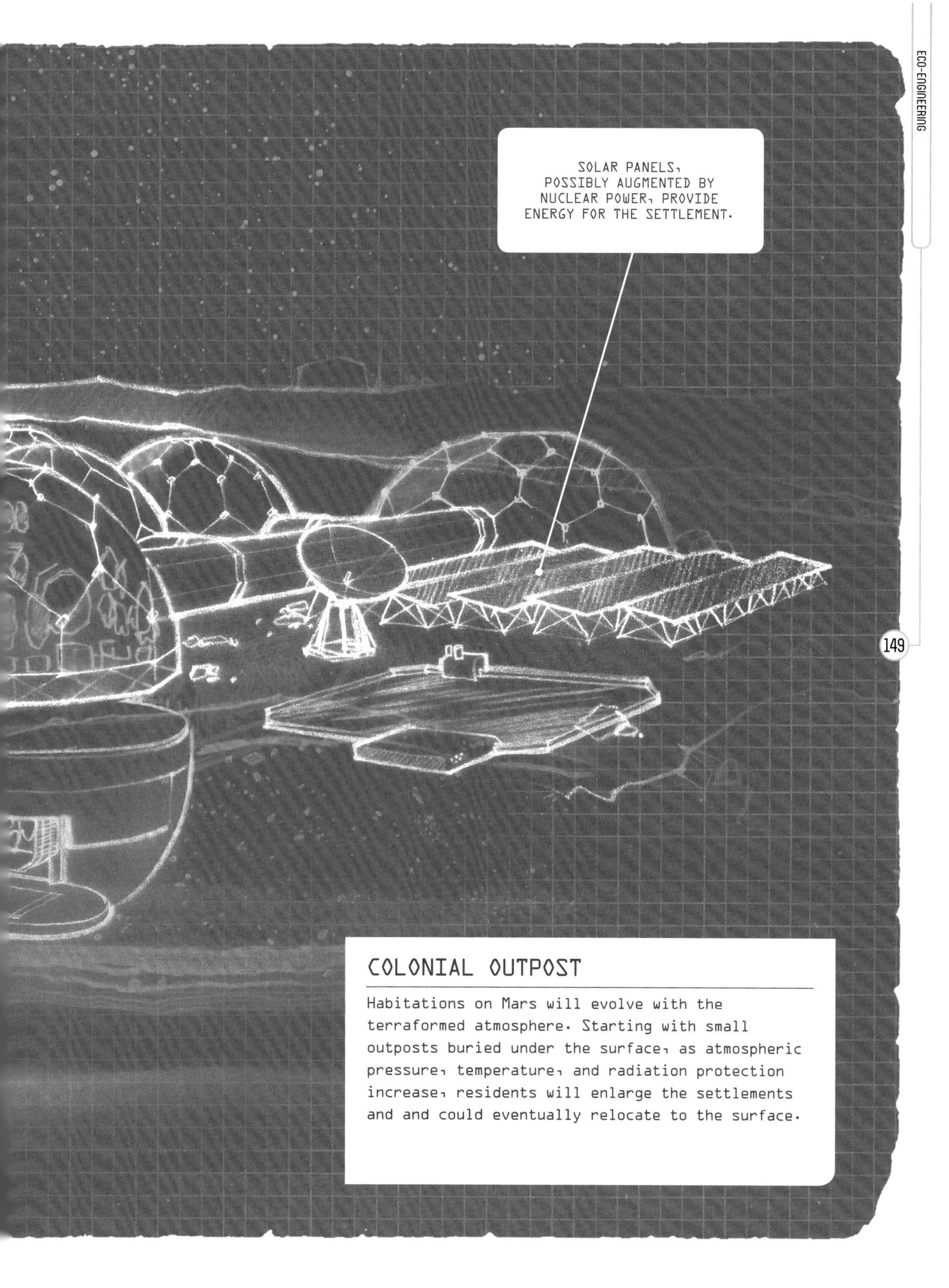

COLONIAL OUTPOST

Habitations on Mars will evolve with the terraformed atmosphere. Starting with small outposts buried under the surface, as atmospheric pressure, temperature, and radiation protection increase, residents will enlarge the settlements and and could eventually relocate to the surface.

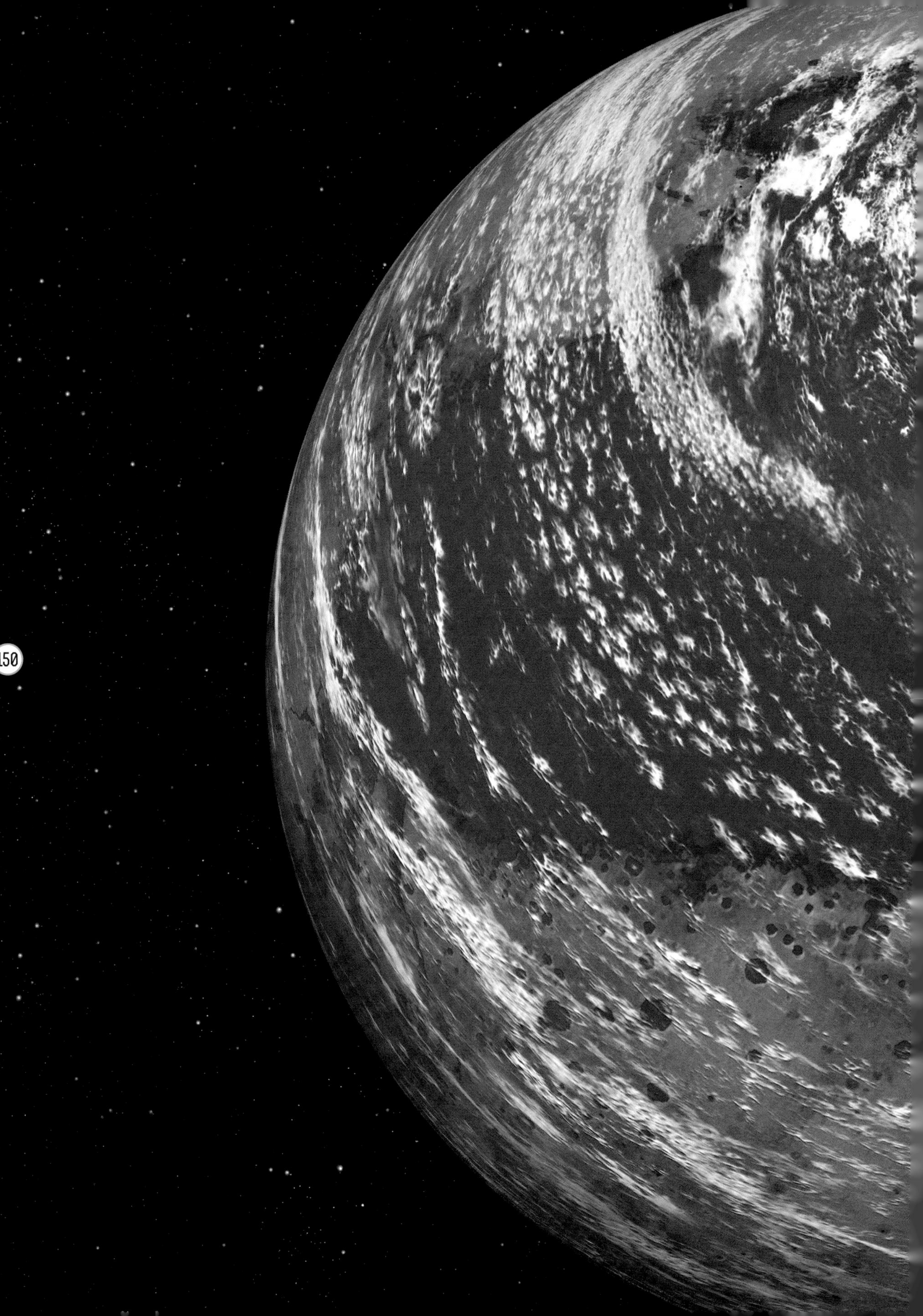

BLUE MARS:
THIS ARTIST'S DEPICTION SHOWS MARS AFTER SEVERAL CENTURIES OF TERRAFORMING. THE NORTH POLE APPEARS RELTIVELY UNCHANGED, WHILE MUCH OF THE NORTHERN REGION HAS RETURNED TO THE TEMPERATE OCEAN THAT ONCE DOMINATED IT. CLOUDS OF WATER VAPOR HAVE BEGUN TO RETURN TO THE MARTIAN SKIES, THANKS TO INCREASING ATMOSPHERIC PRESSURE AND TEMPERATURES.

PLUGGING IN: BIO-PORTS AND CRANIAL INTERFACES

Perception is a tricky thing. *The Matrix* was a chilling but effective meditation on our perception of reality, and how it might not be exactly what we think it is. The movie looked at escaping from alternate reality, but it is more realistic—and more comforting—to assume that what we see *here* is the reality and to develop electronic devices to alter our perspective and to 'enter' an alternate world.

QUESTIONING REALITY

There has been a surprising amount of abstract scientific work done on questions posed by movies like *The Matrix*. The key one, however, is how do we know we aren't living in a simulation? On one side of the argument is that measurements of certain fixed physical properties—the characteristics of cosmic-ray particles, for instance—come from outside our 'reality' and might not measure properly if we lived in a simulation. Others claim that a a sufficiently high-fidelity simulation could account for this. There are other arguments, but none has proved to be entirely bulletproof and a simulation system might have other safeguards built that would make the truth unknowable.

What might a real-life near-term 'matrix' interconnectedness be like, and how close we are to creating—or entering—one? In the movie, a computer programmer named Neo realizes that he lives in a simulated reality—his life is not what he thinks it is. In his reality, all humanity lives inside the matrix, a machine-ruled simulated reality in which our bodies are kept inside little cocoons. They harvest our body heat and electrical activity, and in return give us nourishment and a perception of a 'normal' life. Once Neo stumbles upon this truth, he becomes the target of the bad guys

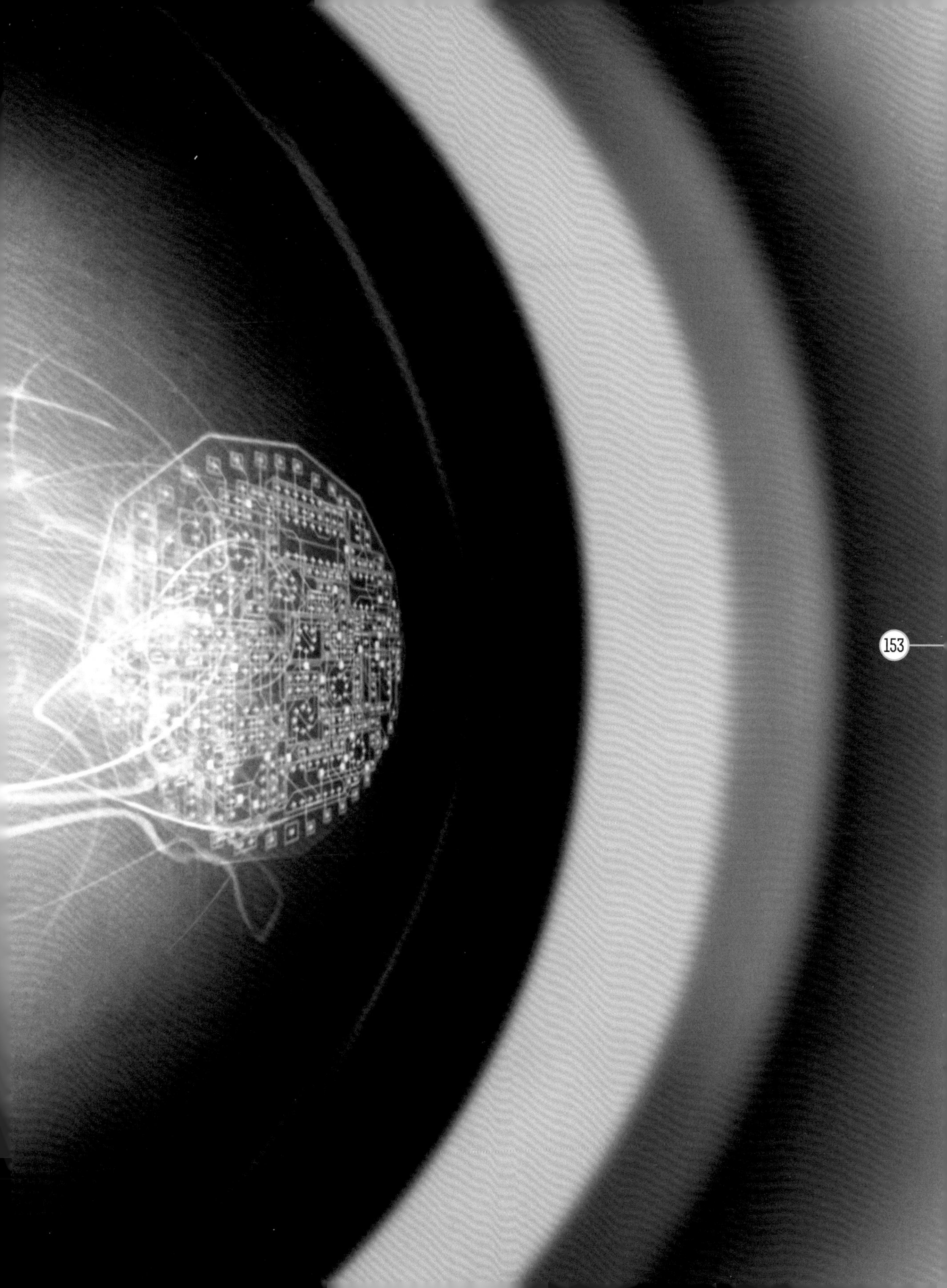

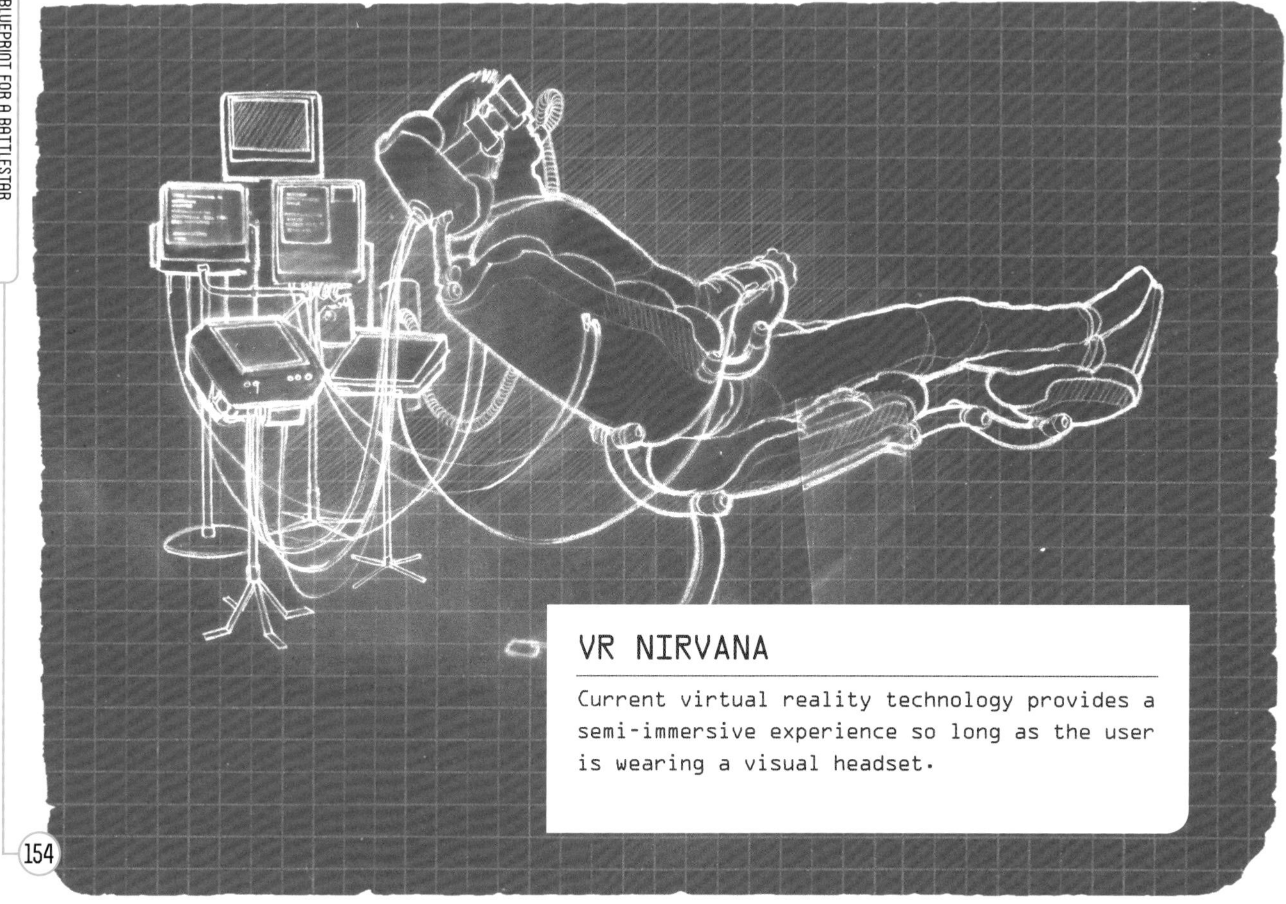

VR NIRVANA

Current virtual reality technology provides a semi-immersive experience so long as the user is wearing a visual headset.

who are manifestations of the Matrix-making machines. Neo finally escapes to freedom in humanity's last refuge, Zion.

In our world—at least, the one we think we live in—the virtual worlds we can experience are a bit less impressive, but also a lot less nastily invasive. How, close, though are we coming to a Matrix-like existence? While humanity is not yet immersed into a fully alternate reality, our day-to-day lives have changed dramatically due to the effects of the 'matrix' that surrounds us now: the Internet and its many manifestations.

Think of the amount of shopping you probably do online, and the (sometimes) helpful push-advertising that vendors send you, after crunching the data they have captured on your interests via your browsing and ordering habits. Think about how we gather information now, from looking at restaurant menus to researching books and other online purchases. Compared to the way things were done just twenty-five years ago, it's a stunning change.

Today's Internet grew out of DARPANET, started by the United States, Britain, and France a few decades back to link defense installations and academic and research institutions with a robust, fault-tolerant computer-to-computer network. Since then it has grown out of recognition, to a decentralized network encompassing 60 trillion individual pages. Online behemoths like Google have done a masterful job of keeping all this material cataloged with web-crawlers, constantly sniffing out new and changed information. These are sorted and indexed by content and key phrases. Algorithms have been developed to convert all this data into searchable results, returned when you ask a question on a search page.

This amazing system enables us to seek out anything our hearts desire, but it goes well beyond personally directed searches. Banking, telecommunications, entertainment, medicine, and military applications are all being handled via various layers of the Internet, and this expansion continues with abandon. We have become so dependent upon it, that an interruption to the Internet would seem catastrophic.

In daily use, the common bottleneck seems to be the human/computer interface. Over time this has evolved from bulky computer workstations with small, bubble-shaped screens and clunky keyboards, to refined versions of these such as laptops, and finally to the smartphones and tablets which are now common methods of accessing the Internet.

But as computers get smaller and faster, the things that have not shrunk along with the chips are batteries and the human-computer interface. The latter still depends on fingers and keyboards or virtual keyboards on screens. The solution? A type of neural or brain-computer interface (BCI). We have already looked at virtual reality interfaces, but BCI goes one big step further: You may recall the brain-implant limb-control experiments we discussed when looking at cyborgs. Research has continued in this direction, but with the goal of offering complete neural control over computer and online interfaces.

Early work concentrated on 'brain caps,' bulky skull-caps with dozens of wires that lead to rack-mounted computers to process the signals into something useful. But vast advances in miniaturization and processing power have resulted in far smaller processing units, some the size of a donut, with sensors that implant directly into the brain. These are obviously not yet ready for the casual user, and are directed more at people with limited body function who need assistance signaling external devices. Newer interfaces, such as 'Cereplex-W' from Brown University, can pick up and interpret the incredibly faint signals from neurons inside the brain and transmit about 48 megabits of data per second, about the same as the average home Internet connection, to an external processor. The system uses tiny silica needles inserted into brain tissue to read neural messages, and is powered by a small, long-lasting battery, bringing us a step closer to a compact and practical BCI.

The Cereplex-W also crosses one other huge threshold, which has long been the nemesis of such devices—the ability to safely bridge the gap from inside

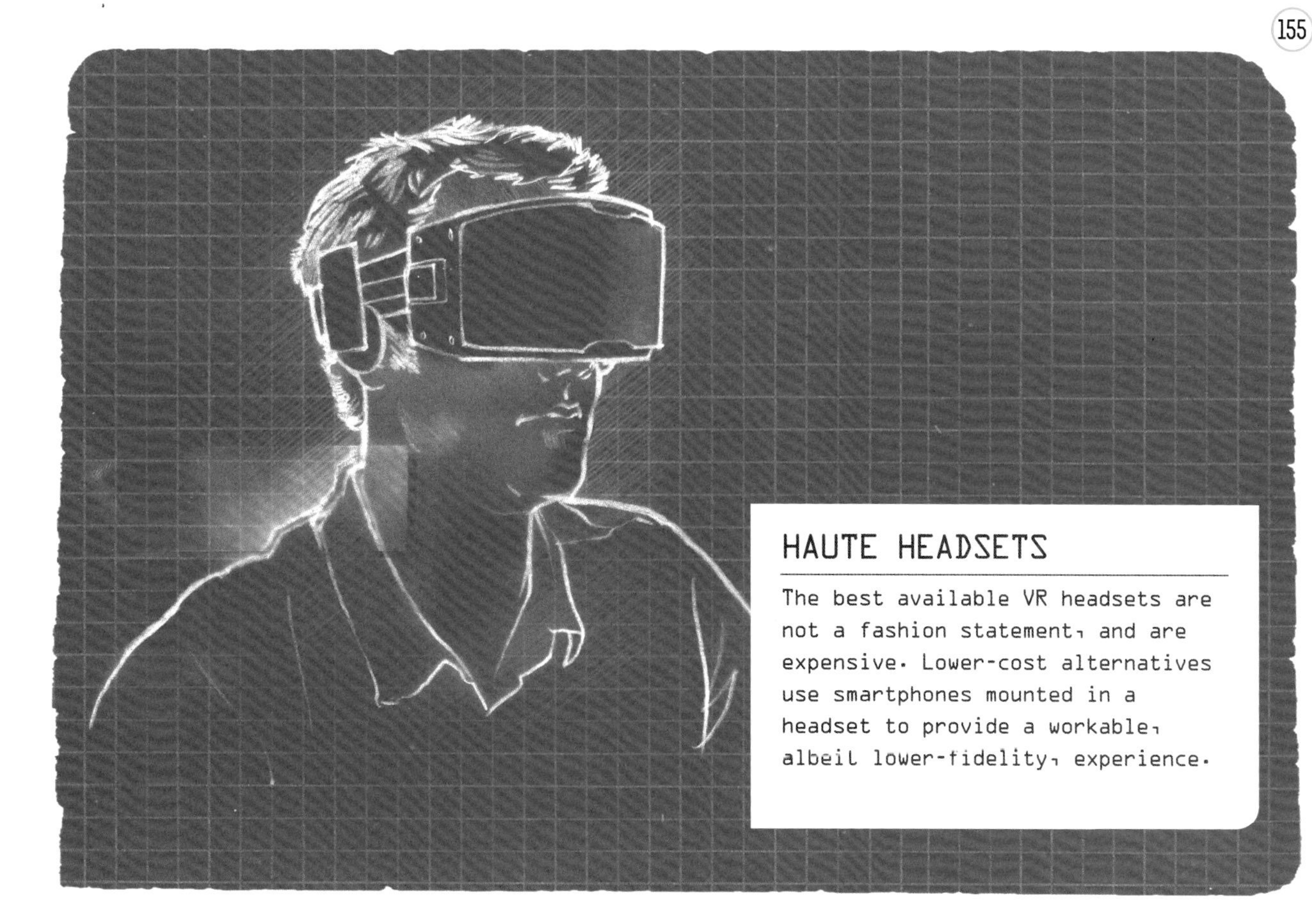

HAUTE HEADSETS

The best available VR headsets are not a fashion statement, and are expensive. Lower-cost alternatives use smartphones mounted in a headset to provide a workable, albeit lower-fidelity, experience.

the body to the devices outside (running permanent wires through the skin is dangerous even in a medical setting). The Cereplex-W is implanted within the head—sensors, processor, and battery—and transmits to the outside wirelessly. Such engineering could one day provide the perfect man-machine interface. Given the current rate of miniaturization and processing power, and research underway into more powerful, smaller processors that create less waste heat, the entire contraption could end up smaller than a pea. We still need to deal with the power requirements, and bullky batteries are one further roadblock. But power may be pulled from an induction charger outside the body that charges through the skin, or may soon be scavenged from inside the body via heat and chemistry.

It is worth mention that many things that are easy for humans—such as talking and walking—are very hard for robots, whereas much that is challenging for many of us—math, manipulating millions of numbers in calculations—is very easy for computers. It's clear our "hardware" has been optimized for survival-related activities, which is why learning to run with a missing leg, while requiring some effort, is doable, as is learning a new language. Human brains are very good at interpolation and learning.

Other research with small implants has shown stunning results in less expected areas. Researchers at DARPA have succeeded in experimentally treating everything from PTSD to depression to restoring lost memories, all via electrical implant stimulation inside the brain. Parkinson's disease can be managed, and reportedly even functional IQ can be enhanced with the same technology.

Human-computer interfaces still have to overcome one really big problem, however. The brain is a hugely powerful processor, operating at *petabyte*-speed. (A petabyte is a trillion megabytes; one megabyte is eight megabits; megabytes are used to measure storage and megabits speed, but you get the idea.) This awesome brain power is enough to run a full-blown computer interface and certainly a web browser. Those interfaces need to become a whole lot more robust to cope with Matrix-style data loads.

Externally worn brainwave readers are making progress, as well. Experimental devices have enabled subjects to communicate words to each other without speaking and across great distances. This is still done via sensors-to-wires-to-processors, which transmit data to the Internet, which is delivered to another set of processors, then to yet another set of external brain electrodes attached to the receiving party. But it is amazing work (and one of the Holy Grails of the gaming industry, which is putting substantial financial investment into this area).

Imagine the power of an externally worn BCI, when used with the Internet, game, or imaging software, and a good virtual reality device like the Oculus Rift headset. Such a system would create an effortless, convincing simulation with instantaneous feedback and control. Making it seamless is some way off—having to wear a big box strapped to your eyes is a giveaway this is not reality—but given time even that difficulty may be overcome.

One other approach to virtual reality is a development of the old MRI scanners that patients used to crawl into for a whole-body scan. fMRI or functional Magnetic Resonance Imaging, is a newer version of those old MRI scanners that can read patterns within the brain from the outside by tracking blood flow. By recording what happens in a given part of the brain during a particular kind of decision or emotion, and building a database from this, fMRI machines can do a pretty good job of determining the user's intent. For now, they are bulky and expensive, but if headphone-sized fMRI machines look like a money-making proposition, the investment in them or something functionally similar but better will materialize.

Other related systems—some of which are nearly ready for prime-time use—track heart rate, facial expression, eye pupil dilation, and, of course, external tracking of brainwaves to accomplish basic interface selections, but these are primitive in comparison to what can be gleaned from inside the brain itself. Think of it as a button versus a bank of sliders—simple selections such as move→click are achievable, but more complex interactions require more complicated tracking of intent.

The systems that read brain signals still require conscious, focused effort on the user's part and, usually, some level of training. And they are still bulky and invasive. For now, nobody can steal your thoughts unobserved. But given time, who knows?

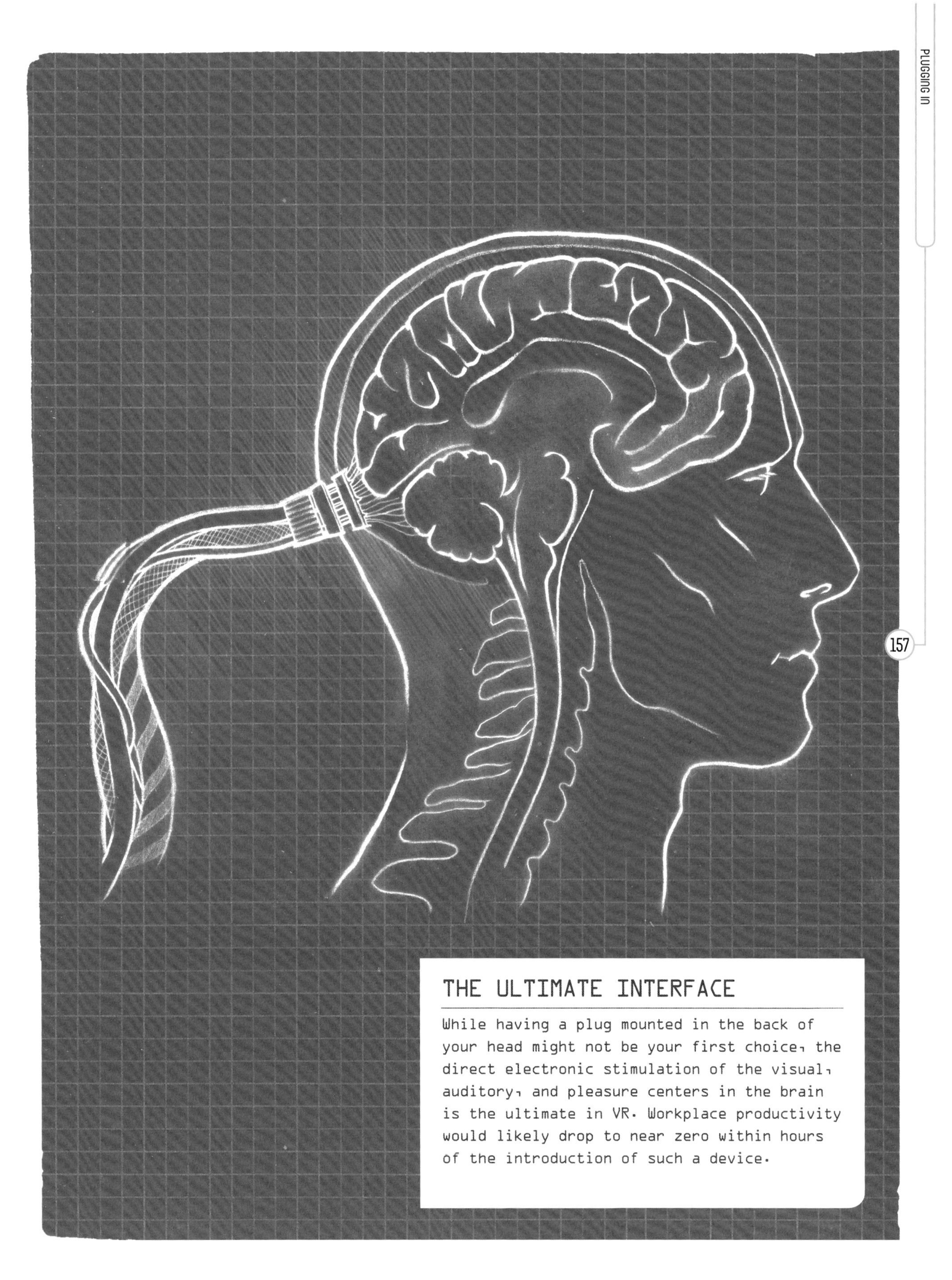

THE ULTIMATE INTERFACE

While having a plug mounted in the back of your head might not be your first choice, the direct electronic stimulation of the visual, auditory, and pleasure centers in the brain is the ultimate in VR. Workplace productivity would likely drop to near zero within hours of the introduction of such a device.

WHO'S OUT THERE?: FINDING EXTRA-TERRESTRIAL LIFE

For decades now we have been searching the skies for signals from another star. To some, this is an exercise in scientific logic—*someone* must be out there. To others, the search for extraterrestrial intelligence (or SETI) is an almost religious pursuit. For still others, it seems like a waste of resources. And, for a well-informed and slightly paranoid few, it's a dangerous exercise in the interstellar equivalent of ringing the dinner bell.

INVASION SCARES

Invasions from outer space have been part of the collective conscious at least since Voltaire wrote of giants visiting Earth from Saturn and Sirius in 1752's *Micromegas*. But it was H.G. Wells' *The War of the Worlds*, published in 1897, that aroused deep-rooted fears of alien invasion. Not coincidentally, the idea of *benign* extraterrestrials was contemporary to Wells' martian tale. In that same decade, Percival Lowell wrote of an advanced and peaceful Martian civilization, capable of a global world government, which was working furiously against the ravages of nature to save their dying world.

RADIO ASTRONOMY DISHES, SUCH AS THE ONES SEEN HERE AT THE VERY LARGE ARRAY IN NEW MEXICO, SEARCH THE HEAVENS FOR POSSIBLE MESSAGES OF EXTRATERRESTRIAL ORIGIN.

CLUSTERS OF IDENTICAL NANOPROBES FLY IN ROUGH FORMATION TO REACH DISTANT STARS. SOME WILL MALFUNCTION AND OTHERS WILL LIKELY BE DAMAGED, BUT ENOUGH WILL REACH THEIR DESTINATION FOR A SUCCESSFUL OUTCOME.

CONTACT

The first human ambassadors to other stars may take the form of swarms of nanoprobes. The can range from DVD-sized down to the size of a postage stamp. They will have transmitters to send data back to Earth, and will be powered by giant lasers on or in orbit around Earth. Using this method of propulsion, which negates the need to carry fuel on-board, these small 'sails' can reach incredible speeds in short order.

There are, of course, those who believe that we were contacted by extraterrestrials thousands of years ago. For decades, people have proposed that 'Arcturans' may have shown the ancient Egyptians how to build the pyramids using long-lost levitating machines. Andromedans are said to have drawn the Nazca lines in Peru and Aldeberaneans are claimed to have visited the prophet Ezekiel (as some people seem to interpret his Biblical writings).

Little of this bears scientific scrutiny, and we have probably been alone since amphibians first crawled out of the primordial sea. But if fextraterrestrial beings never visited us, where are they?

This is not a new question. Enrico Fermi, who helped build the first atom bomb, pondered it. His thinking was verbalized in 1950 when, over a luncheon with his colleagues, he posited the Fermi Paradox. This states that the reason we haven't been visited by aliens yet might be that interstellar flight is impossible; or, if it is possible, has been judged to be not worth the effort; or that technological civilization does not last long enough for it to happen. To be fair, the Paradox was passed on by another diner at that lunch, and we cannot be sure if Fermi was questioning the existence of extraterrestrial civilizations (as is widely believed) or simply the ability (or interest) of those civilizations to reach us.

But the question remains: our solar systems is comparatively young, about 4.6 billion years, or just under one-third the age of the universe. So, given that there are a) billions of other galaxies of all ages, and b) stars with planets within them, and c) that there could be intelligent life on some of these with the ability to develop interstellar or even intergalactic travel technologies—why haven't we seen them? Presumably some have mastered the technology required, so where are they?

Even if these aliens have been unable (or, perhaps wisely, unwilling) to pay us a visit, why haven't we detected them via other means? Humanity has been blasting television shows into the cosmos for decades and early 20th century radio programming would be detectable well over 100 light years away by now. So assuming that other technological civilizations use electromagnetic transmissions for communication (a fair assumption, though still just that—an assumption), why haven't we detected them?

Not long after Fermi first posed his paradox, an American astrophysicist named Frank Drake turned up the heat. In 1961, prompted by some other scientists thinking over the lack of any sign of extraterrestrial life, he prepared a compelling argument for a conference on pursuing SETI via radio telescopes, a relatively new concept at the time. He devised an equation based on a number of broad assumptions, which was really intended as an exercise to provoke thought on the subject, not to provide the final word.

In very general terms, it includes:

- The number of civilizations in our galaxy which should be emitting electromagnetic emissions that we can measure (N)
- The rate of star formation friendly to intelligent life (R_*)
- The fraction of those stars (the above) with planets (f_p)
- Number of the above planets that support that life (n_e)
- The fraction of the above planets on which life appears (f_l)
- The fraction of the above planets on which intelligent life emerges (f_i)
- The fraction of the above civilizations that develop suitable communication technology (f_c)
- The length of time those signals are detectable (L)

So, $N = R_* \bullet f_p \bullet n_e \bullet f_l \bullet f_i \bullet f_c \bullet L$

The calculation depends on a number of variables for which it is necessary to make big suppositions. As a result, different attempts to calculate N have come up with findings ranging from optimistic to almost despairing. Drake's own calculations yielded an estimate of ten civilizations in our galaxy that we should be hearing from. As we learn more about exoplanets and the adaptable nature of life to extreme environments on Earth, some of these values should become more valid and the results therefore somewhat more trustworthy. But even if the answer is ten (or five, or 30), where should we be looking and for what do we listen? For those involved in SETI, that has been the big question for years.

Nikola Tesla suggested in the late 1890s that a powerful version of radio, which he had invented (though the credit usually goes to Marconi), might be able to establish contact with intelligent beings on Mars. Marconi and other radio pioneers picked up on this, but little was actually done toward the goal.

Then, in 1924 there was a close approach of Mars (it edges closer to Earth every two years, but some of these approaches are closer than others; this one was the closest for almost a century). The US government called for a 'National Radio Silence Day.' All radio stations would go silent for five minutes every hour for three days. A radio antenna was lofted in a dirigible (a rigid blimp), but after 36 hours nothing special had been detected.

Mars stepped out of the limelight shortly thereafter, and by 1959 scientists were thinking in larger terms. Why limit yourself to Mars? Frank Drake started a project in 1960 called 'Ozma' (after the princess in the Wizard of Oz books). He pointed a radio telescope located in Green Bank, West Virginia, at the star systems of Tau Ceti (which as you will recall, is only about 12 light years away) and Epsilon Eridani (around 11 light years away), but found little other than the usual background noise. The Soviet Union took an interest and conducted their own experiments, which resulted in the same outcome.

Some universities tinkered with the idea, but it was in 1971 that SETI got serious. NASA finally devoted funds to commission a study that resulted in an ambitious design for a program called 'Project Cyclops,' which would have built radio telescope arrays with multiple smaller dishes instead of vastly more expensive large dishes. But as with so many NASA studies, it was a project built only on paper; SETI would have to wait another decade.

Ohio State University had been quietly conducting its own SETI search in the same time period, but jumped into the media fray in 1977 when a bizarre signal was received by its radio dish. A SETI researcher had steered the university's radio telescope toward a deep-space object called M55, a

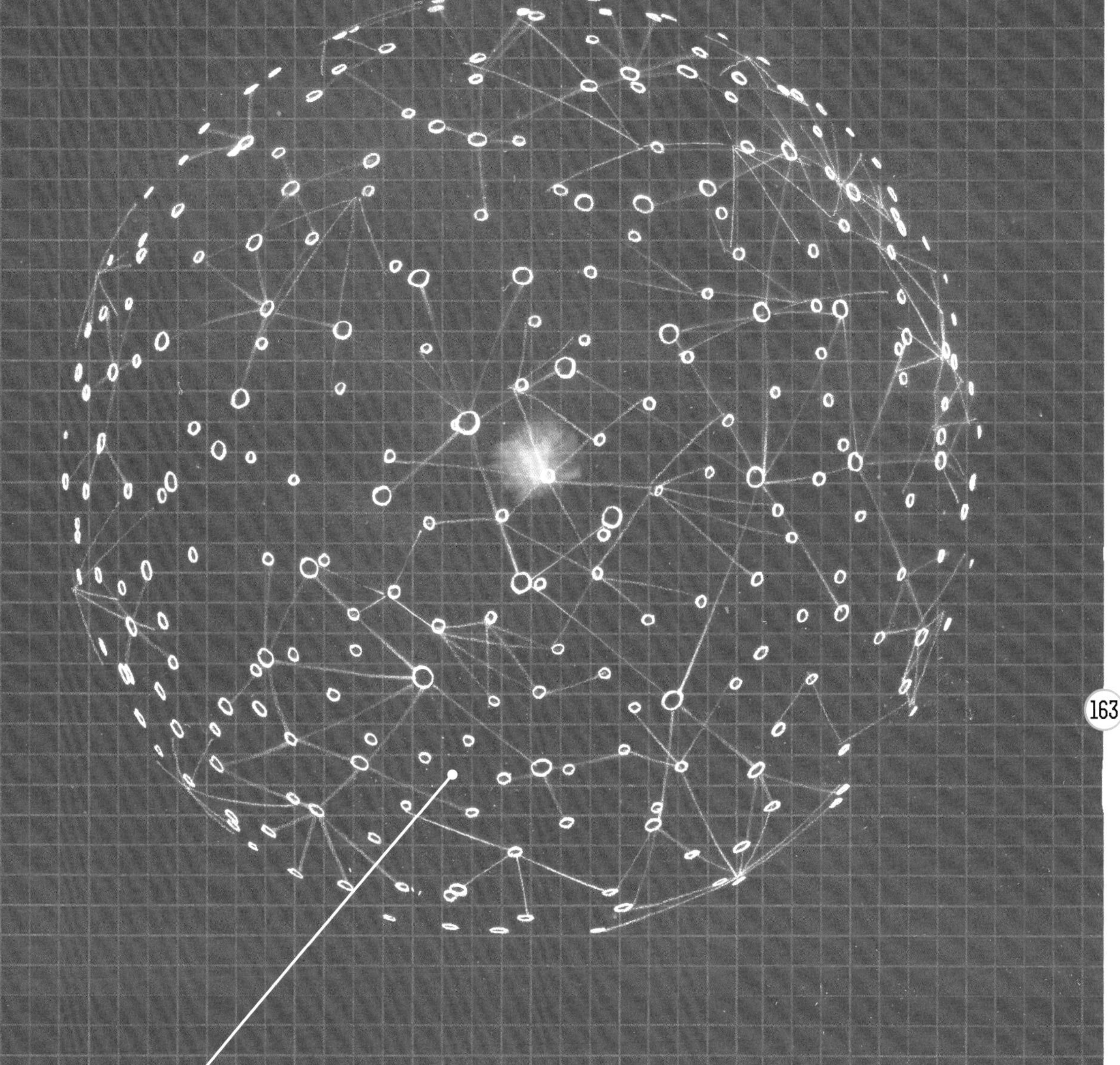

INDIVIDUAL POWER COLLECTION SATELLITES WOULD REMAIN IN POSITION AND UTILIZE COLLISION-AVOIDANCE TECHNIQUES TO MAINTAIN A STABLE NETWORK FOR POWER COLLECTION AND TRANSMISSION.

MR. DYSON, I PRESUME

In 1960, famed physicist Freeman Dyson postulated advanced civilizations might meet their ever-increasing energy needs by directly collecting the energy coming from their star. A Dyson sphere, or shell, would be a solid globe surrounding the star, and a Dyson Swarm, seen here, would be an open-lattice or freely orbiting collective of energy collection units.

globular cluster—a closely packed spherical grouping of stars inside our galaxy. On August 15, 1977, an unusually strong signal was received in the hydrogen band of radio frequencies, which was deemed to be ideal for long-range radio transmissions. For just over a minute, the signal continued, until M55 moved out of range due to Earth's rotation.

The Ohio radio telescope measured radio signals in ten-second intervals and assigned a numeric code to each. Suddenly, out of the hash of normal background noise, a sharp spike in signal intensity occurred, smoothly ascending and then descending in value as the dish swept past whatever was emitting the signal. The researcher was startled, and excitedly (and famously, in SETI circles) drew a red oval over the paper printout, and wrote a big red 'WOW!' next to it. It has been known since then as the 'WOW Signal.'

Despite returning to that region of the sky more than 50 times since, scientists have never detected a similar signal again. It was apparently a one-time burst of energy. Theories on its origins ranged from a classified military satellite to Earth-based transmissions bouncing off space debris. Or some kind of bizarre star-emission. Or, of course, aliens. But nothing similar has been heard since.

In 2016, another alternative suggestion was put forth by a professor at a Florida college. He noted that there were two comets in the right place to have crossed the aperture of the radio telescope on the date the 'WOW Signal' was intercepted—and comets travel with large hydrogen clouds that could have been a signal source.

Modern SETI took root when the Planetary Society was formed in 1980 by Carl Sagan and a group of other space scientists from NASA and JPL. Using more advanced digital signal processors, they

designed a new SETI system that would search far more broadly, and also be able to sort nearby signals from more distant ones. This project has continued in one form or another since, but with no confirmed results to show for their trouble. It was begun with NASA funding, but over time the US Congress carved away the funding until most large-scale SETI work in the US died of economic starvation.

Then an offshoot of Sagan's efforts, the SETI Institute, was founded in 1984. With an impressive roster of fine minds at the helm, and some private funding, the non-profit has been operating continuously since then. NASA and the National Science Foundation have kicked in funds as well—it's easier for the government agencies to add to an existing pot of cash. When we see news about SETI today, it is usually associated with the institute. Using a variety of both radio and optical telescopes, the SETI Institute conducts organized, regular searches of the sky for new signals.

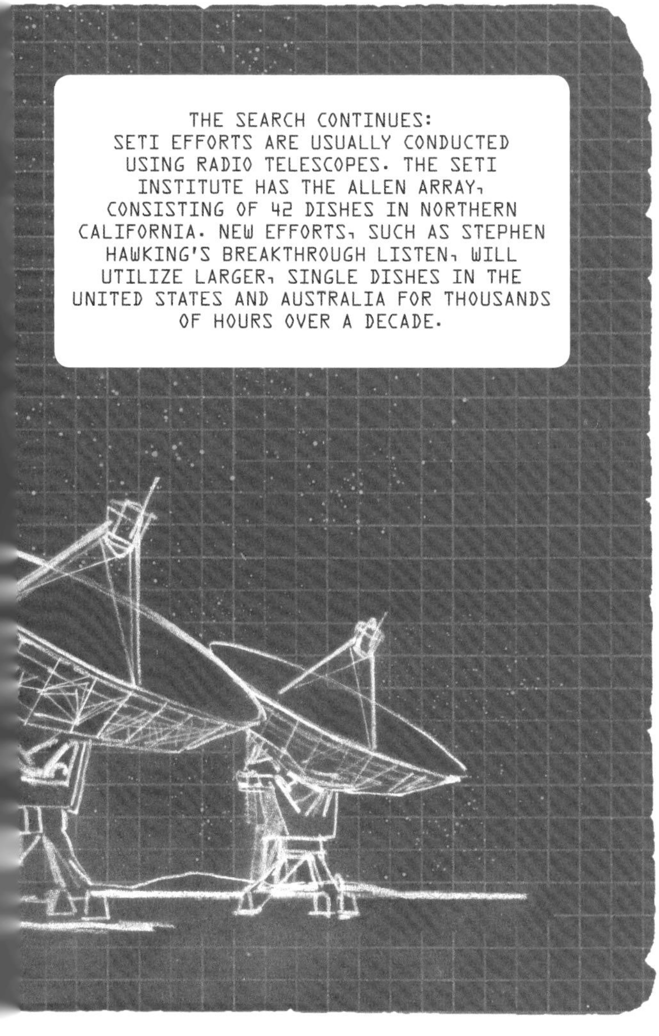

One more major SETI effort has joined the hunt for cosmic intelligences. Called Breakthrough Listen, funded by Russian billionaire Yuri Milner and supported by Stephen Hawking, it is a ten-year, $100 million effort that will survey the skies with radio telescopes around the world, using newer tools and at a still broader array of frequencies. The survey will operate for thousands of hours per year as opposed to the dozens that other efforts have had the funding to accomplish, and cover ten times as much extraterrestrial real estate. The combined effort is said to be sensitive enough to pick up a signal equivalent to commercial aircraft radar from the 1,000 nearest stars.

After all these decades of listening, an obvious thought is to *send out* a message to be received by others. This has been attempted on various occasions, but SETI is considering a larger, orchestrated effort. Called 'Active SETI,' the idea seems like a winner...

But not so fast, say some, including Stephen Hawking. For some years, the famed physicist has warned that we need to think long and hard before reaching across the void to our galactic neighbors. He, and others of a similar mind, feel that there are very real risks associated with bringing ourselves to the attention of aliens, who may have technology far superior to our own and might not be friendly. They could invade for reasons of need, economic gain, or simply because humans taste good.

SETI still operates in passive mode, sneaking radio peeks at thousands of likely targets throughout the galaxy. Soon, we may send out a loud interstellar hello. By the time it's likely to be intercepted, we should have galactic battle cruisers of our own, and any trouble can hopefully be averted.

The search for extraterrestrial intelligence continues. Some astronomers feel strongly that the evidence, scant though it is, indicates that we should hear something within a decade or two...assuming, that is, there is anyone to hear from. The universe is a very large, very old, place and it's rather difficult to believe that there are no other intelligent species out there. Let's find them...before they find us.

THE SEARCH CONTINUES: SETI EFFORTS ARE USUALLY CONDUCTED USING RADIO TELESCOPES. THE SETI INSTITUTE HAS THE ALLEN ARRAY, CONSISTING OF 42 DISHES IN NORTHERN CALIFORNIA. NEW EFFORTS, SUCH AS STEPHEN HAWKING'S BREAKTHROUGH LISTEN, WILL UTILIZE LARGER, SINGLE DISHES IN THE UNITED STATES AND AUSTRALIA FOR THOUSANDS OF HOURS OVER A DECADE.

STOPPED COLD: BECOMING IMMORTAL

Immortality is a practical impossibility, at least in our human form. Something *approaching* immortality may be possible once we reach the singularity and have our minds uploaded. But at that point, who will be tending the servers?. For our corporeal forms, it appears that we will have to be content with extreme longevity. It does appear that aging, while a natural process, does not have to occur at the rate that it does. We might as well be living to 140 instead of an average of 80 years or so.

PROGRAMMED AGING

The reasons that aging kills appear to be a combination of damage to DNA from external factors, and the human body's own internal clock, such as the shortening of telomeres (which sit at the end of DNA strands, protecting the DNA from damage when cells divide, but themselves get shorter at each cell division.) Over time, this shortening encourages reproductive defects, leading to disease and organ failure.

Programmed aging is the body's way of telling you that it's time to remove yourself from the gene pool. In general, we seem to enter decline shortly after

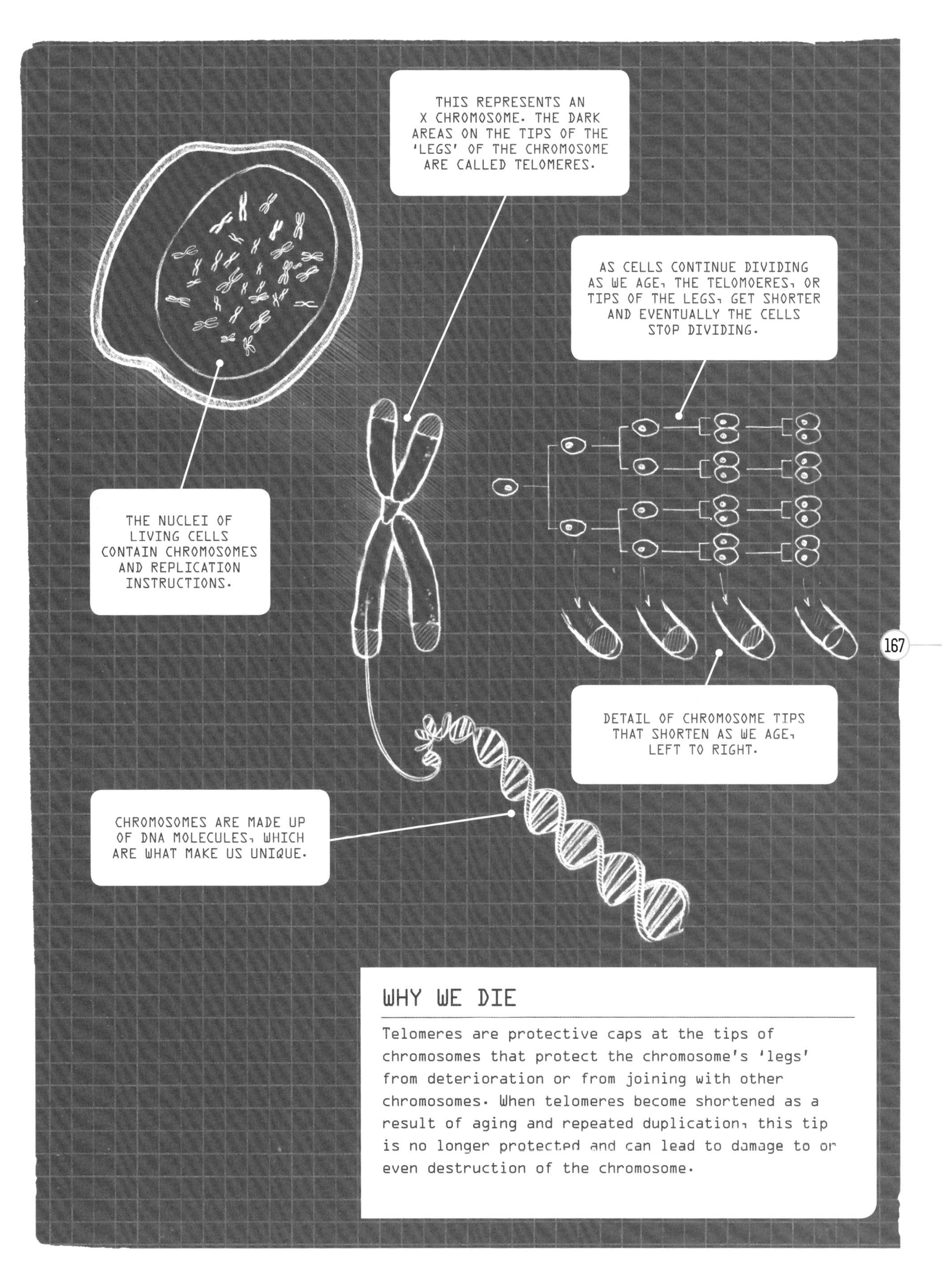

WHY WE DIE

Telomeres are protective caps at the tips of chromosomes that protect the chromosome's 'legs' from deterioration or from joining with other chromosomes. When telomeres become shortened as a result of aging and repeated duplication, this tip is no longer protected and can lead to damage to or even destruction of the chromosome.

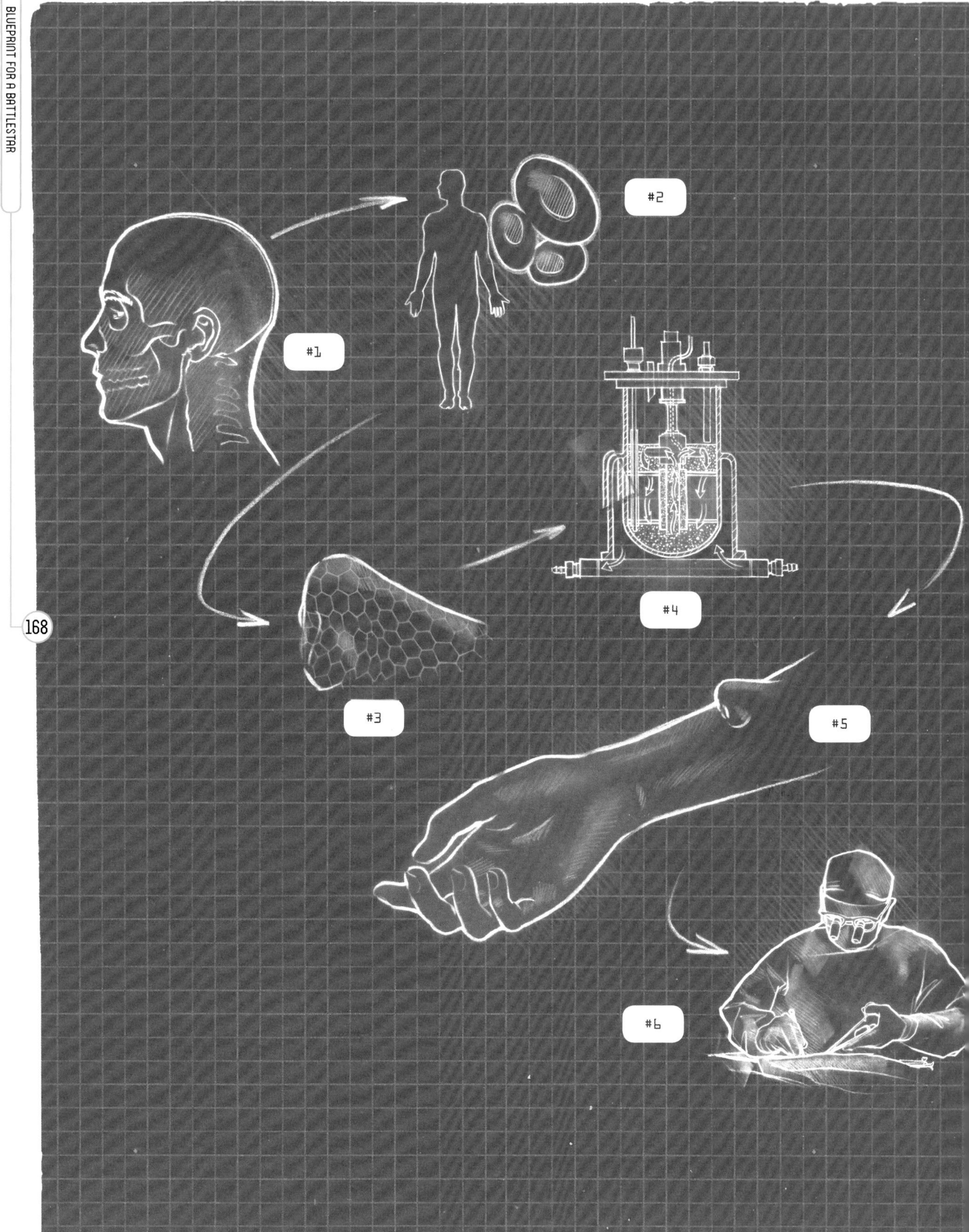
#1
#2
#3
#4
#5
#6

#1: A PATIENT HAS LOST HIS OR HER NOSE TO CANCER OR AN ACCIDENT.

#2: STEM CELLS ARE HARVESTED FROM PATIENT'S BONE MARROW, THEN ALLOWED TO MULTIPLY IN THE LAB.

#3: STEM CELLS CULTIVATED AND SPRAYED ON A CUSTOM MOLD MADE IN THE SHAPE OF THE PATIENT'S OLD (OR IMPROVED) NOSE.

#4: NOSE PLACED IN 'BIOREACTOR'—A SMALL CHEMICAL-FILLED VAT—TO ENCOURAGE GROWTHS OF CARTILAGE.

#5: GROWING NOSE IS SURGICALLY ATTACHED TO PATIENT'S ARM TO CONTINUE GROWTH AND GENERATION OF BLOOD VESSELS AND SKIN.

#6: A FEW MONTHS LATER, NOSE IS ATTACHED TO PATIENT'S FACE.

GROWING NEW ORGANS

Medical science now has the capability to create limited replacement components for our bodies using stem cells.

our prime reproductive years. Shortened telomeres seem to be a part of this scheme—it is as if the body has an internal counter that says, 'At a certain time I'm going to shut off your ability to grow new and viable cells.' There is also a theory that natural selection plays a part in this; that certain mutations that favor fertility in youth can cost you on the other end of your life.

Environmental, or external, damage is the other leading contender for shortened life spans. This can take place via diet, exposure to certain chemicals, radiation (including sunlight), stress, and more. On average, about a third of cancers resulting from DNA damage seems to be caused by external influences. Most of the rest of the DNA that gets carved up within you may be from viral infections.

Nature, as usual, provides many clues about longevity, but also tosses in a few puzzles. Mayflies have a lifespan of about 24 hours, the world's shortest, while the hydra—a relative of the jellyfish—is capable of self-regeneration and is nearly immortal. For the gardeners among you, annual plants reproduce sexually, but die at the end of the season. Perennials effectively clone themselves and live forever.

Other animals live a long time, and much research has been conducted into the natural causes of this trait. Lobsters can live somewhere between 35-70 years, but nobody is really sure. They are an interesting case, as they appear to become more fertile with age and retain their vitality and energy levels until shortly before death. This may be in part due to the levels of telomerase in their systems, an enzyme that actively repairs their DNA as it ages and begins to deform. In most vertebrates, this enzyme declines rapidly with age.

As far as mammals go, one species of whale, the Bowhead, can live for over 200 years. Researchers looking at the creature's longevity have identified in the animal's DNA up to 80 genes that seem to give it enhanced ability to repair DNA, slow cellular aging, and resist cancer. Lower metabolic rates may help longevity too.

Age-cheating genes from our fellow mammals could be very helpful, as there may end up being a way to introduce whale genes into humans who need specific types of tissue or organ rejuvenation. This could involve creating lines of transgenic stem cells with both human and whale DNA that, when introduced to the human body, will be able to migrate where needed and repair tissue damage.

So what near-term medical advancements might we expect to enable the extension of our life spans? The above-mentioned genetic interventions could be the single most revolutionary shift of all, since it now appears that our genetic makeup—and the alteration of it—is a major culprit in aging. So whale genes may have a place in your future, along with other stem cell blends.

Recent research into the telomerase enzyme, so powerful in lobsters, has also proven very promising toward restoring a youthful state. Mice that had been deprived of the enzyme, and showed signs of severe aging, were then given large doses of it and regained their health quickly. The researchers concluded that the enzyme could be effective against a number of age-related disorders, or even that the treatment could slow aging *before* those disorders had a chance to manifest themselves.

So far we have looked at systemic ways to slow or reverse the effects of aging. Of course, as we learned in Chapter 16, we will soon be able to become cyborgs and transcend our biological forms. We are not there yet, but replacement organs are one step in that direction.

For decades, surgeons have been able to replace some organs when they begin to fail. But the procedures have continued to have problems. For one thing, you need to find a replacement organ that is compatible with your physiology. And the older and less healthy you are, the less likely you are to be matched with an appropriate replacement in a timely fashion.

Xenotransplantation is a procedure that has been experimented with for many years, but until recently has shown little signs of being successful in the long-term. This is the process of placing organs from certain animal species—such as pigs and primates—into humans. Naturally, there are a number of issues involved. For one thing, most of the compatible donor animals have shorter life spans than human beings, so the organs tend to age

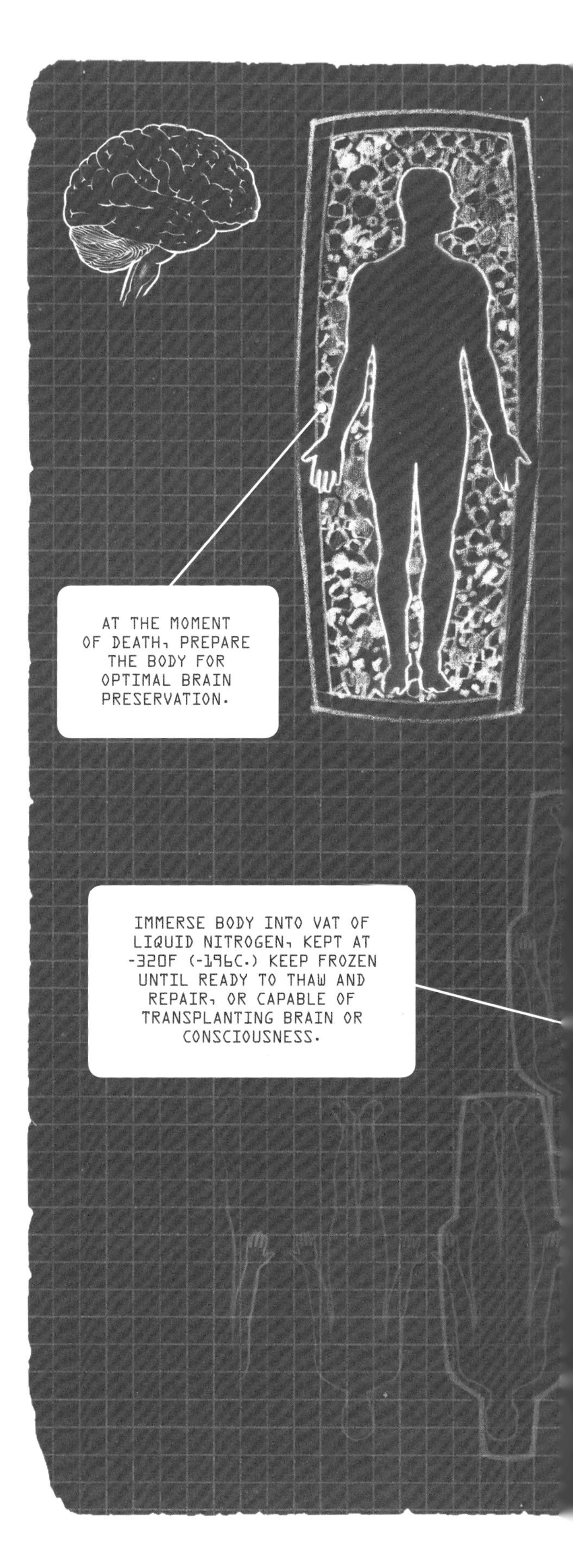

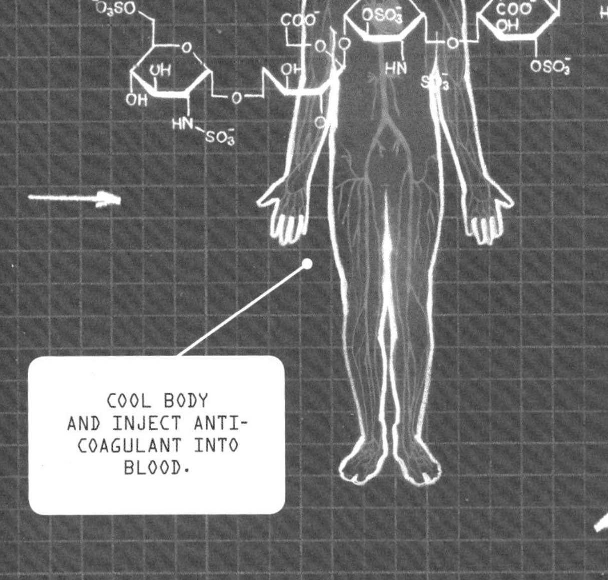

CHEMICAL ANTIFREEZE IS USED TO REPLACE WATER IN THE BODY, DOWN TO THE CELLULAR LEVEL. COOLING CONTINUES DOWN TO ABOUT -200F (-129C.)

NEVER GIVE UP

At the moment of death, we may one day be able to preserve the body for later resuscitation by carefully freezing it, preserving the physical structure, and the brain—all that makes 'you' you.

faster than their human counterpart. But if you are 65 and looking for a new part, that may not be such an issue. The larger problem is that of rejection. The human immune system is not happy when a piece of another animal is placed inside the body and reacts to the affront. Hyperacute rejection is the immediate rejection of the new organ or tissue, and results in severe inflammation and often the death of the implant. Acute vascular rejection takes more time—a matter of days—but the results are the same. Chronic rejection takes much longer, but causes a hardening and narrowing in the circulatory system, and again, the organ fails. Research is underway to better understand and discourage these reactions, but there is still a long way to go.

The ultimate replacement for a failing organ would appear to be via regrowth of part or all of the organ. The best way to go about this is to use stem cells, the human body's universal 'fixit' part. The good news is that embryonic stem cells from a human can be used to repair or replace most parts of the patient's body. Otherwise they must come from embryos. This has led to enormous ethical complications, a battle that rages on in many countries. There can also be rejection issues with stem cells from another body and immunosuppressant drugs must often be used to keep the immune system from performing a search-and-destroy mission on the implanted stem cells. Experiments to overcome this by implanting the nuclei of the patient's cells into the incoming stem cells have shown promise, but it appears that the best way to recreate needed tissues from stem cells is to use the patient's *own* stem cells. Recent advances have indicated that adult stem cells, while not quite the universal panacea that embryonic cells are, can do much more than we gave them credit for. In the near future, this form of therapy may become the magic bullet we have been seeking for dozens of ailments.

There have even been successful experiments done to transform adult skin cells into their original state. Since all humans begin existence as a mass of stem cells that only later specialize and become a specific organ or tissue, this could, in theory, ultimately lead to growing whatever new part you need from a skin scraping. The tough part of this is 'reprogramming' the cell to become what you want it to, and this is just now beginning to be understood. Again, there is a long way to go before you will be able to buy a take-home kit and grow your own body parts, but this might be achievable in time.

Immortality is more complex than simply dipping your ruby-encrusted chalice into the Fountain of Youth, as the conquistador Ponce de Leon is reputed to have tried to do in 16th-century Florida. Taking care of the body you have and all its constituent parts, and living one evolutionarily dictated lifetime of 95 years, may prove to be enough for most people.

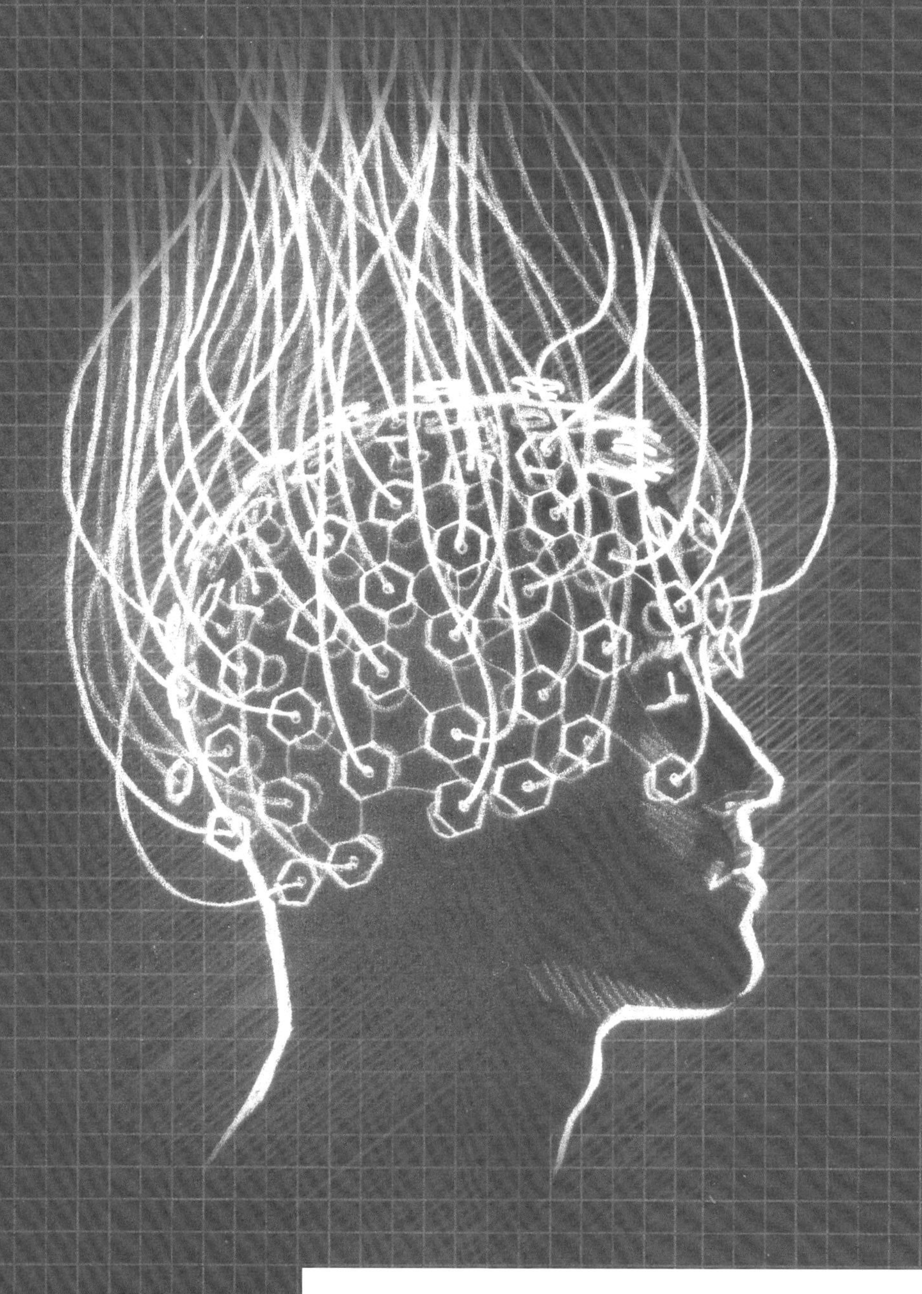

TRUE IMMORTALITY

Probably the best guarantee of living forever would be to become a computer yourself and transfer your consciousness into a data network. But can this eventually be made to work, and will it be true consciousness or just a collection of memories and impulses?

DISAPPEARING ACT: ACHIEVING INVISIBILITY

Being invisible is a fantasy, like flying under your own power—for now, experienced only in our dreams. We have all craved invisibility at one time or another, for our own (often nefarious) reasons. Whatever your motivations, it's actually not too far off. Not close enough to sneak past the the security barrier at your favorite concert, but there is some fantastic work being done, and the pace of research is accelerating. In real life, invisibility is like the ultimate form of camouflage. You are attempting to blend in with your surrounding environment to the point that you cannot be seen, or at least not easily. This can be accomplished by melding with the colors and lighting of that environment, by reflecting or projecting part of that environment back to the viewer, or by finding a way to redirect light so that you have no reflection at all. Of course, we all want the third option, which is akin to Harry Potter's cloak of invisibility that we drape over ourselves and then go sneaking around.

THE ULTIMATE CAMOUFLAGE

Like so many things of military value, invisibility has evolved greatly over the past century. As late as the end of the 19th century, the world's militaries were still marching around in splendidly colorful uniforms—the British had their red tunics, the Germans varying shades of battlefield blue with red stripes, and the US Marines were still dressing like cavalry soldiers in pale blue. All very pretty, but not exactly made to blend into the foliage surrounding them. By World War I, this had all changed. Uniforms, in general, were khaki brown, and camouflage had become the order of the day: ships,

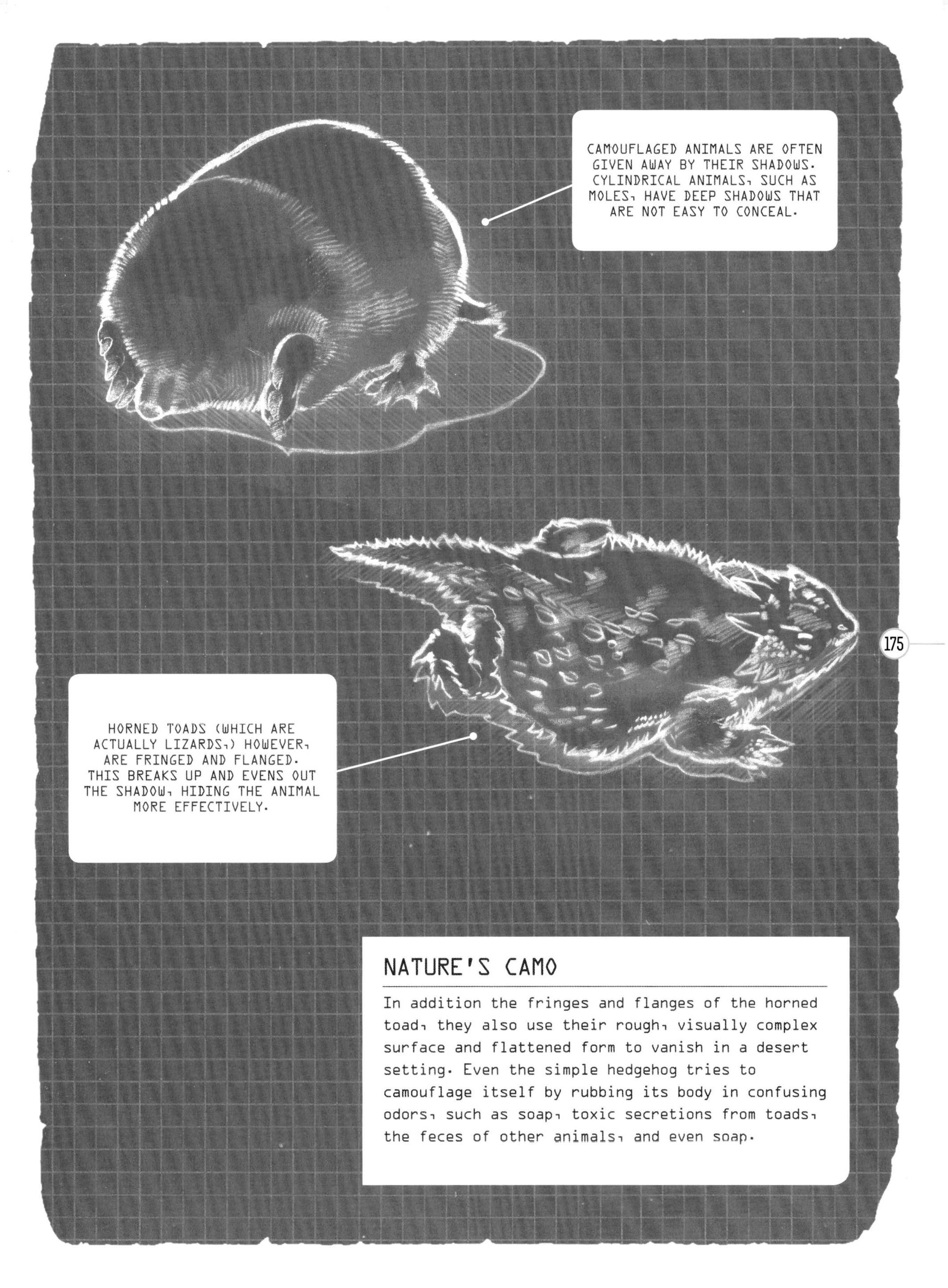

NATURE'S CAMO

In addition the fringes and flanges of the horned toad, they also use their rough, visually complex surface and flattened form to vanish in a desert setting. Even the simple hedgehog tries to camouflage itself by rubbing its body in confusing odors, such as soap, toxic secretions from toads, the feces of other animals, and even soap.

tanks, and artillery positions were finally disguised to merge into the background, as were the soldiers manning them.

Humans were finally learning from nature, which had been perfecting camouflage for eons. We have all seen the moths that blend into tree bark so well as to be effectively invisible. Similarly with stick bugs—rather than just adapting to the color and patterning of their environment, these insects adopt both the color and form of their surroundings, looking indistinguishable from a twig.

Chameleons are a very particular case. Some have the ability to change their skin color and patterning. The flesh of these chameleons contains guanine crystals, and over this, a layer of skin with color pigments. As the chameleon reacts to a stimulus by changing the spacing between these crystals, the wavelength of light reflected from their skin changes. These changes can be due not only to the need to hide, but also to temperature, emotional state, and aggression. These reptiles use what is known as active camouflage, something that humans are just figuring out how to do.

Color is just one way to fade into an environment. Confusing the eye of the observer with patterning (usually in conjunction with color blending) is like hiding in plain sight. Tree moths are a good example: when properly aligned, they vanish into tree bark perfectly. But sometimes the goal is more about confusing the eye than blending in.

In World War I, ships were at constant threat from enemy submarines. An idea called dazzle camouflage was hit upon by the British and Americans about the same time, inspired either by a zoologist or an artist, depending on whom you believe. Once again, nature provided clues—zebra stripes and other animal markings provided inspiration. The idea was not to try and hide a ship in mid-ocean, which is no small feat, but rather to make it as difficult as possible to estimate its size, speed, and direction, making the calculation of an effective torpedo attack against them tough. A number of schemes were tried—most involved large, daring slashes of dark and light stripes intersecting at odd angles. Although the coloration and patterning actually made the ships more visible as

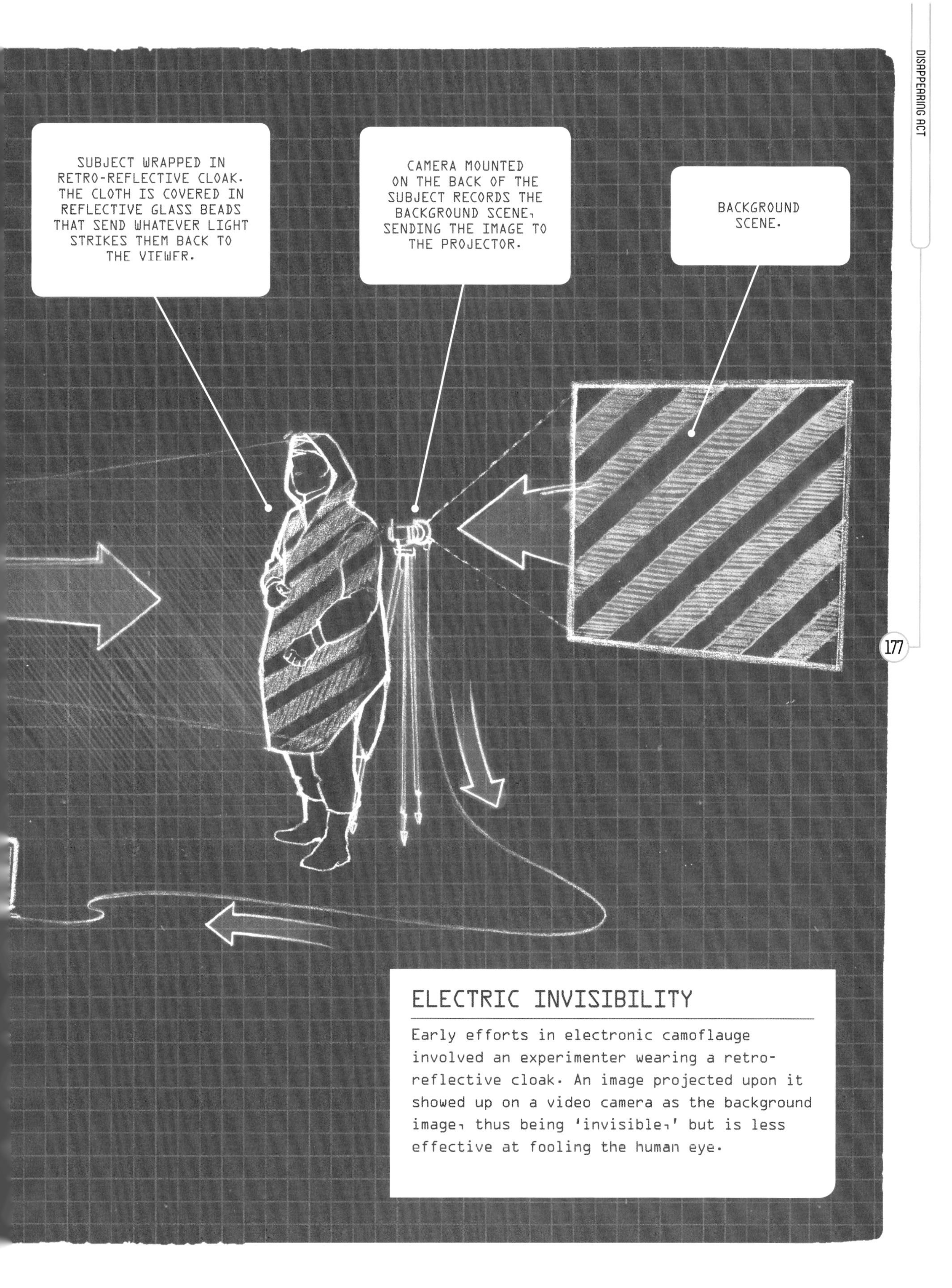

ELECTRIC INVISIBILITY

Early efforts in electronic camoflauge involved an experimenter wearing a retro-reflective cloak. An image projected upon it showed up on a video camera as the background image, thus being 'invisible,' but is less effective at fooling the human eye.

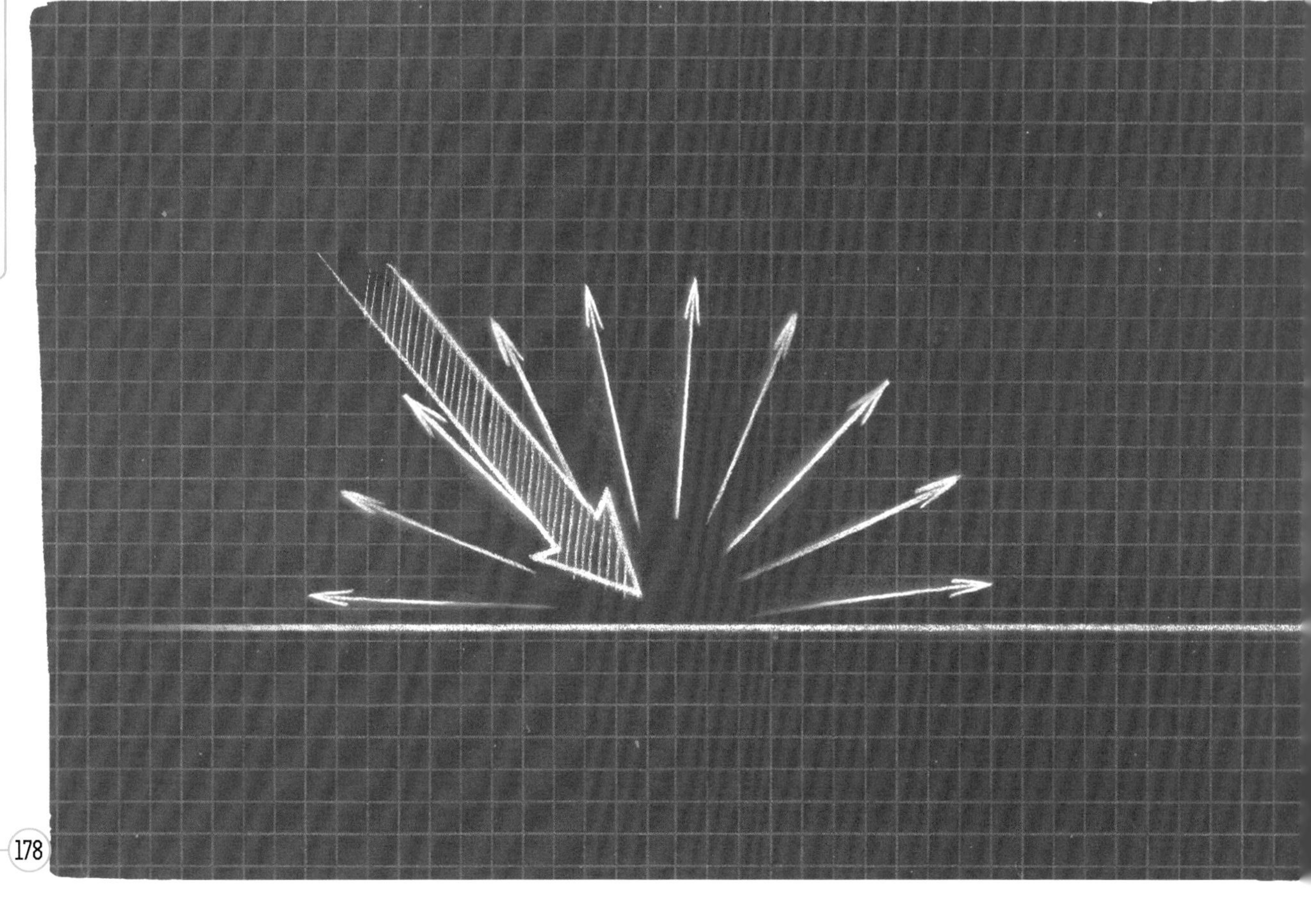

opposed to less, it made it difficult for submarine captains to judge where the ship was heading, and where it would be when their torpedoes arrived. Dazzle schemes had some success until radar came along and spotted ships using radio waves no matter how they appeared visually.

Blending and dazzle are examples of 'passive camouflage.' So are the military uniforms mentioned earlier. The US military changed years ago from the green-patterned combat fatigues to brown ones, as their missions were redirected from the forests of Europe and Vietnam to the deserts of the Middle East. Later revisions changed the blobby black patterns on the clothing to pixelated little squares, known as 'digital' camouflage, designed to better blend into the surroundings via the principle of fractals—the design works at different scales. This pattern is in effect a scaled version of itself, repeated across the clothing, and is more effective at a broader range of distances than the earlier designs were.

The large foliage-blending camo-nets that are routinely spread over artillery and other bulky combat emplacements, utilize blending on a couple of levels: first, color, and pattern is chosen to look as much like the surrounding environment as possible; then changes in lighting, which create ever-changing shadows, are minimized by reducing shadows. Once again, nature provides examples of this, with insects that are shaped to hug the ground, minimizing drop-shadows from the sun overhead. It is a complex and tricky business, and we are fortunate that the bugs took a few million years to work it out for us.

Not all camouflage is designed to fool the human eye; sometimes it is radio waves that are the foe, as with radar. This is where stealth technology comes into play, as used on the US Air Force's B-2 bomber and F-117 aircraft, as well as some ships. These combat machines are designed with shapes and edges that scatter and confuse enemy radar, as well as coatings that absorb radar waves. The aircraft

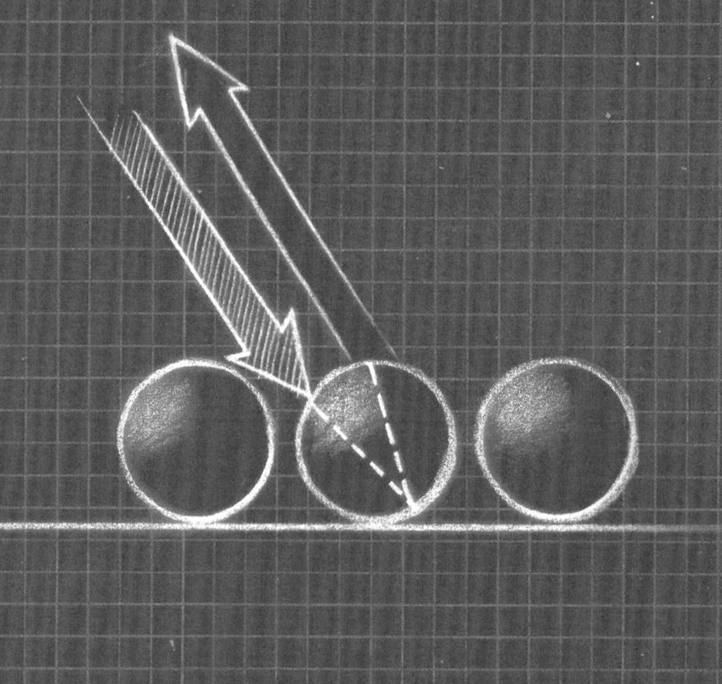

GOING RETRO

Retro-reflective materials are covered with tiny glass beads. When a ray of light hits the bead, as on right, the beam is bounced right back at the viewer, creating a duplicate of a projected image. To left, a light ray striking a normal surface is simply scattered. These kinds of reflective mechanisms only work in certain circumstances.

also place their engines in such a way that the heat emanating from them is invisible to infrared tracking; their 'heat signature' is obscured. Research is proceeding on combat uniforms that will do the same for soldiers, effectively shielding the heat from their bodies from night-vision scopes.

But none of this amounts to an invisibility cloak. How can we simply become truly *invisible*?The answer to that is more involved, and requires batteries—a lot of them. But once electricity gets into the game, things do get much more interesting.

A few years back, someone got the idea of hiding behind a television. Sounds odd, but when the TV was showing a live shot of what was behind it (and behind the person being obscured,) it was a somewhat convincing illusion. But what if you could project that background image onto a reflective cloth? That is more effective, and in 2003 a professor at the University of Tokyo experimented with this. It works on the same principles as the TV screen illusion, but uses instead a special cloth that is retro-reflective—that is, it reflects back the image that is projected onto it.

The fabric is covered with thousands of glass beads. When an image is projected onto it, the beads, being spherical, bounce the light inside themselves, then back out toward the projector. If the person looking at the cloth is standing close to the projector, he sees a return of the projected image, in effect rendering the cloaked subject 'invisible.' A refinement of this would require the fabric to be self-illuminating, containing thousands of tiny video projectors instead of merely reflecting what is shined upon it. Various labs have tried, but it's a tough assignment, made especially difficult by the constantly warping, twisting shape of the fabric.

A more advanced version, created by DARPA and already in use by US SWAT teams (and doubtless the military as well,) uses image-generating fabric that can create a low-fidelity image on the camo-suit, with no projectors needed. The fabric is created with a photo-electric plastic base that can alter color and patterns to adapt to the environment—a tech not dissimilar to that in your e-book reader, which has little magnetic balls that are dark on one side and light on the other. They flip as directed by a magnetic charge. A more advanced version can actually update these changes as the wearer moves. Imaging instructions can come from a body-mounted camera that 'sees' what is behind the wearer, and the suit mimics that image; or more general input from ground-based cameras or even satellite imagery can be used—either can send a general blending pattern to the fabric, which it matches.

Variations of electronic active camouflage have shown promise on larger surfaces. In one system developed in the UK, armored vehicles (like tanks) have video screens attached to their large, flat sides. Images from the terrain on the far side of the tank are fed to the video displays, and in the proper lighting conditions, the images displayed on the screens effectively hide the bulk of the tank. It's not perfect, and parts of the tank still show, but it does hide the overall profile of the tank and makes it hard to identify what exactly it is.

Much closer to an 'invisibility cloak,' is the creation of a group at the Lawrence Berkeley National Laboratory: a type of cloth that belongs to a larger family of fabrics and films called 'meta-materials.' The goal is to create a perfectly reflective suit that, in effect, 'hides' from the light reaching the viewer's eyes. The prototype uses an incredibly thin film that is covered with a varying pattern of tiny brick-shaped gold nuggets. The 'fabric' is 50 nanometers thick and the gold chunks are 30 nanometers thick. A human hair is about 100,000 nanometers wide, so the film is operating at an almost molecular scale. The tiny sizes are a requirement; the reflective elements need to be within the same size range as the wavelength of energy they are trying to manipulate, and when you are talking about light, that's a huge challenge.

In normal situations, the scattering of electromagnetic energy—in this case, visible light—is what makes objects visible to the eye. It is also what makes objects visible to energy detectors, such as infrared cameras or even X-ray detectors, when those energies are reflected by the object being viewed. So normal cloth, skin, or metal reflects, absorbs, or refracts whatever energy hits it, and is therefore viewable. The meta-materials can change the reflective characteristics of a fabric, causing the light rays to curve or even go backward, something that does not occur with normal materials. So, in simple terms, you are taking whatever illumination is being shone upon the fabric, and returning the light back in the direction from which it came. It is like an optically perfect mirror created on a form-fitting fabric. The light waves are not scattered, and none of the colors in visible light are absorbed (which is how we see colors in the first place). So, whatever is cloaked becomes, in effect, invisible. And there is more: by tuning how the tiny gold bricks reflect the light, you can even cause the cloaked object to look like something else—one researcher stated that a tank could be made to look like a bicycle.

This is even harder to accomplish than it sounds, and so far the test subjects are tiny patches of the material in laboratory settings. And at this point, in theory at least, the person wearing the cloak would not be able to see out of it. But the basic technology is sound, and scaling it up to a useful size does not seem like an insurmountable challenge. Once that's done, we really could have a Harry Potter-style invisibility cloak.

Metamaterials

Experiments with metamaterials show that they can provide effective cloaking of 3-D objects. Metamaterials bend the radiation striking the cloaked object, effectively rendering it invisible at certain wavelengths. This technology is new and the experiments are at a small scale, but show great promise for larger-scale cloaking.

BUILDING TOMORROW: THE CITIES OF THE FUTURE

I hope you have enjoyed our trek through some of the remarkable technology of yesterday, today, and tomorrow. While our present—in other words, 'the future' as envisioned in the past—has not turned out quite as we were told it would, it is still pretty remarkable.The technologies of the future will astound. While we have learned that there's a good chance you won't have a jet pack or flying car, there are compensations. You will soon be able to pull out your smartphone—a device that is already more powerful than the whole building full of computers that launched the Apollo Moon rockets—tap your smart car app, and have your Tesla 9000 drive over to you, pick you up, and take you to work before parking itself. Oh, and nanotech and stem cells may cure cancer and replace your defective organs as you age. I'll take that over a flying car any day.

GOING DIGITAL

There are a few trends that have emerged over the past few decades and which are sure to continue. One is the world of bits versus atoms, as posited by futurist Nicholas Negroponte in his 1995 book *Being Digital*. He predicted that everything that could be done digitally would be, and that atoms (i.e., objects) were unwieldy and bits (electrons) were not.

Presciently, this included the paper his book was written on—it is now available as a digital ebook download. But the important thing is that our society has hurtled madly into the world of bits, in a way completely unforeseen even 25 years ago, and this has changed everything. The digital world provides us on demand media, instant information, online

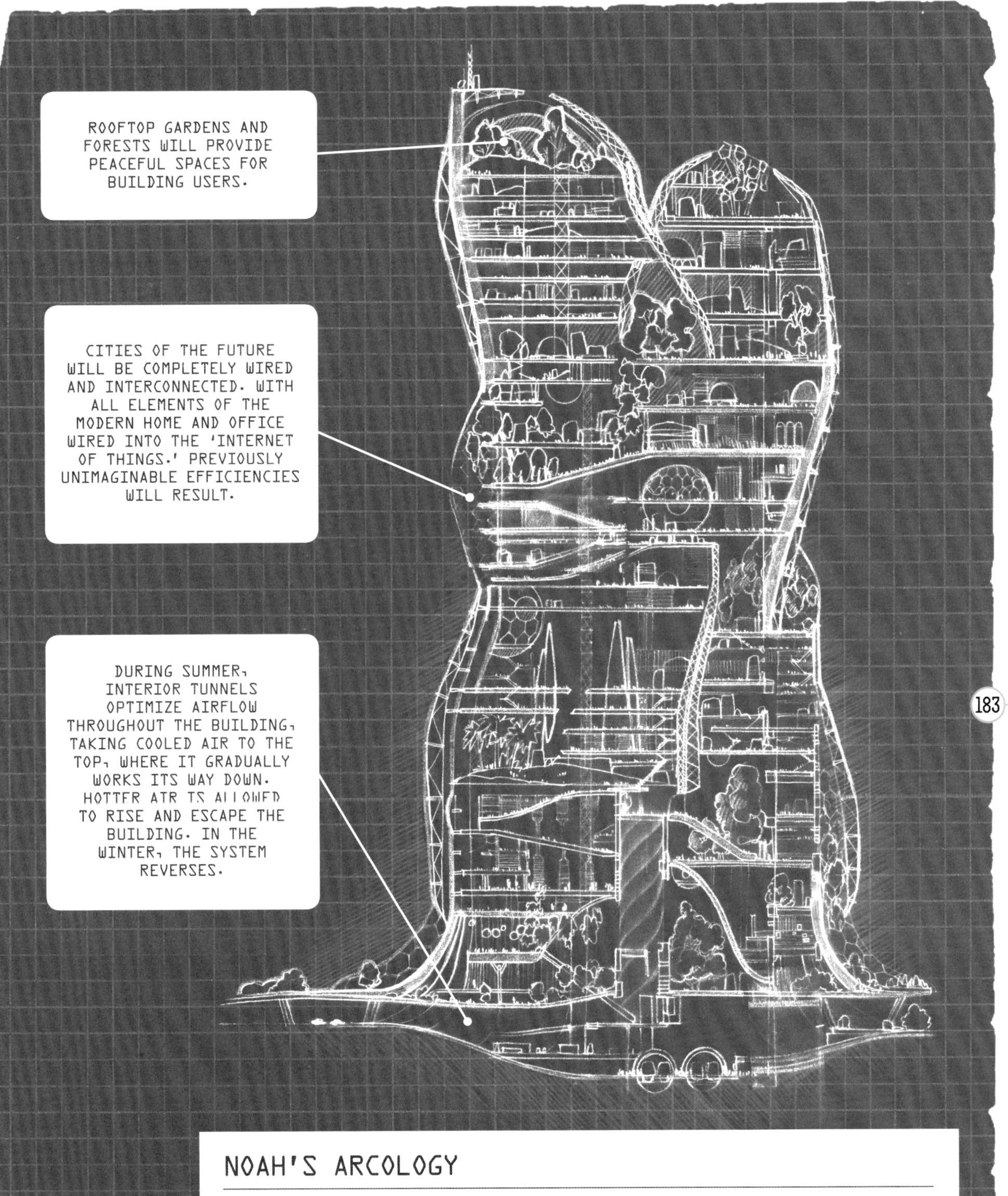

NOAH'S ARCOLOGY

Blending architecture and ecology, the field called arcology strives to optimize structural and environmental design to best serve dense urban areas. By blending the best in new technologies and visual/spatial design, planners can improve living for urbanites while reducing their environmental impact.

shopping, and romance-on-tap. It has also allowed the proliferation of remote-control crime, sucked our children into a world of interactive gaming and virtual-only relationships, and stolen much of our privacy. The present is a mixed blessing. What will the future be like??

Obviously, the move toward the realm of the digital will continue. The explosive growth of the past 25 years may slow a bit in the next decade, as we are reaching the limits of what can be accomplished with existing microprocessor technology (Moore's Law may soon be out the window.) But advances in alternative approaches to manipulating those bits are on the cusp, and quantum computing, the realm of not just 1s and 0s but every possible state in between, is near. Computing will be unrecognizable in a few decades.

Of course, your children will probably be browsing the Interplanetary Internet from their billets in the Mars colony by then. Starships will make stopovers at Mars and Europa before engaging their Alcubierre drives to jump into a warp bubble and head off to other star systems, while Earth will be a clean and verdant natural preserve, protected by international accord from the further depletion of resources.

While you're waiting for SpaceX's Mars Colonial Transporter to pull in, you will get a medical checkup by a completely lifelike robot, then you will lie back, close your eyes, and enjoy your choice of 3-D entertainment in the privacy of your own darkened eyelids, courtesy of your neural implant.

We've touched on each of these subjects, but there's more. What about the larger structure of our society? What will happen to the many have-nots in today's world those so deeply in need of better water, food, and education? Well, there may be good news coming there as well.

Our increasingly scarce natural resources, which are the cause of so many geopolitical squabbles today, should actually become less essential, our endless needs met from other sources. Much of this will be derived from true energy independence; we will move away from messy, toxic, and limited fossil fuels, such as Middle Eastern oil. Our energy will come from the Sun, gathered by orbiting solar power stations that will harvest that abundant and free energy, convert it into a high-intensity signal, and beam it to central distribution nodes on each continent. From there, it will be transmitted along new, smart power grids to where it is needed—everywhere. This limitless source of low-cost energy will allow fresh water to be extracted from the oceans upon demand. Our food supply will likely diverge, with one major source being 'natural' foods—what we now consider to be organic produce and meats, and the other being engineered foods—genetically modified, hydroponic, and artificial nutrients. These will be optimized for health and nutritious content, lowering overall medical costs as people consume healthier foods.

Cities will continue to be the dominant gathering place for people, but they will probably be much more spread out and decentralized. With the web running almost all of our daily lives, proximity will be less of an issue. The abundant energy available will allow us to electrify almost everything we do, from running the home to transportation. And speaking of transportation, traffic jams will be minimized, as self-driving cars optimize routes and speeds to avoid congestion, leading to far fewer accidents due to the computerized elimination of driver error. Ultra high-speed hyperloop-style transport will replace short-hop air travel and the remaining cross-country and international flights will be via hypersonic transport that burns less fuel on the three-hour hop from Los Angeles to Tokyo than a jet airliner does flying from Omaha to New York today.

The arts will flourish to enrich minds freed from much of the time-consuming day-to-day tasks that fill our time now. Self-expression, already enhanced in the early 21st century by software that allows the individual creation of music and visual arts, will also diverge between those who choose to spend their copious spare time with electronic creation versus those who prefer more 'analog' methods of expression, such as acoustic instruments, singing, and painting. Cooking will be more of a creative undertaking than ever, with a ready supply of natural ingredients available as an alternative to instantly available engineered foods.

The individual obsession with wealth will diminish for many people, as more of our basic needs become freely available. The vast majority of people will, for the first time, be able to live lives of abundance—not by spending more, but by doing more of what they love and expressing themselves more fully. Loving relationships will be less stressed by the need to survive, and people will have more time to spend with their families. Crime will drop dramatically, as the needs of more are met with fewer resources and at lower cost. Abundance is the key.

Living and loving will be more comfortable, both for the body and the mind. All but a handful of mental disease will become a minor inconvenience, as truly universal health care provides the world's population with nanotech interventions to prevent or cure the afflicted. Physical disease will be minimized, as new organs are grown from your own tissue to replace old ones as they fail. And they will fail less quickly and less often, as DNA interventions slow aging. What we think of as retirement today will become half of a person's life, enjoyed more fully in good health and ample individual gratification. At least, this is the future I'm hoping for. It is clearly possible, given the scientific and technological advances of the past 20 years.

And what of the darker side of humanity? There will still be crime, war, and poverty, but hopefully less of each. Criminal activity will stay with humanity so long as we have freedom of choice. A few will always choose unsavory shortcuts to fulfill their needs, and quick solutions to what inflames them. But with ample resources for humanity and less emphasis on personal gain, perhaps the drive to acquire wealth beyond what is truly needed—by whatever means—will diminish. In cases where it does not, and some continue to turn to criminal activity for gain, a combination of predictive policing

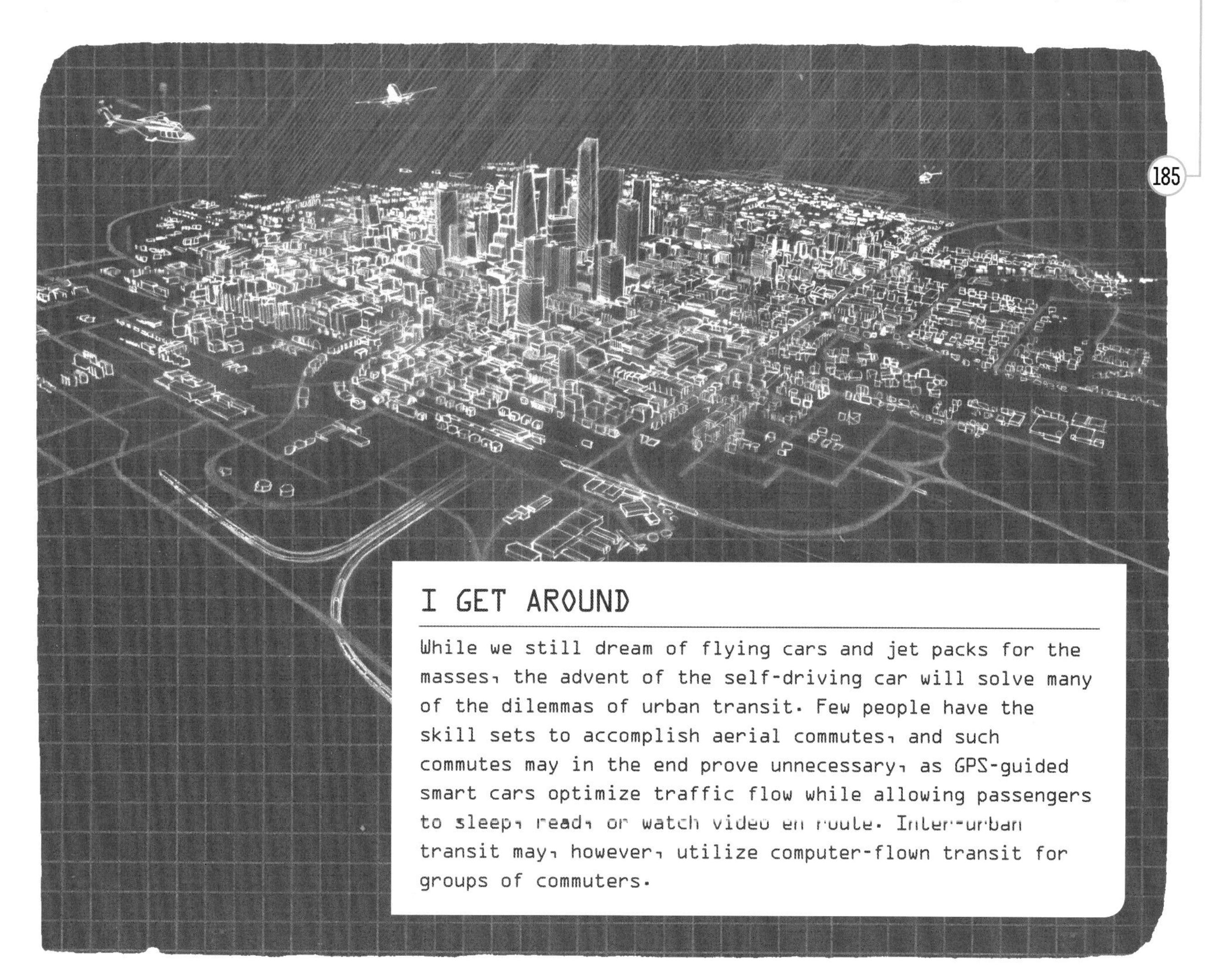

I GET AROUND

While we still dream of flying cars and jet packs for the masses, the advent of the self-driving car will solve many of the dilemmas of urban transit. Few people have the skill sets to accomplish aerial commutes, and such commutes may in the end prove unnecessary, as GPS-guided smart cars optimize traffic flow while allowing passengers to sleep, read, or watch video en route. Inter-urban transit may, however, utilize computer-flown transit for groups of commuters.

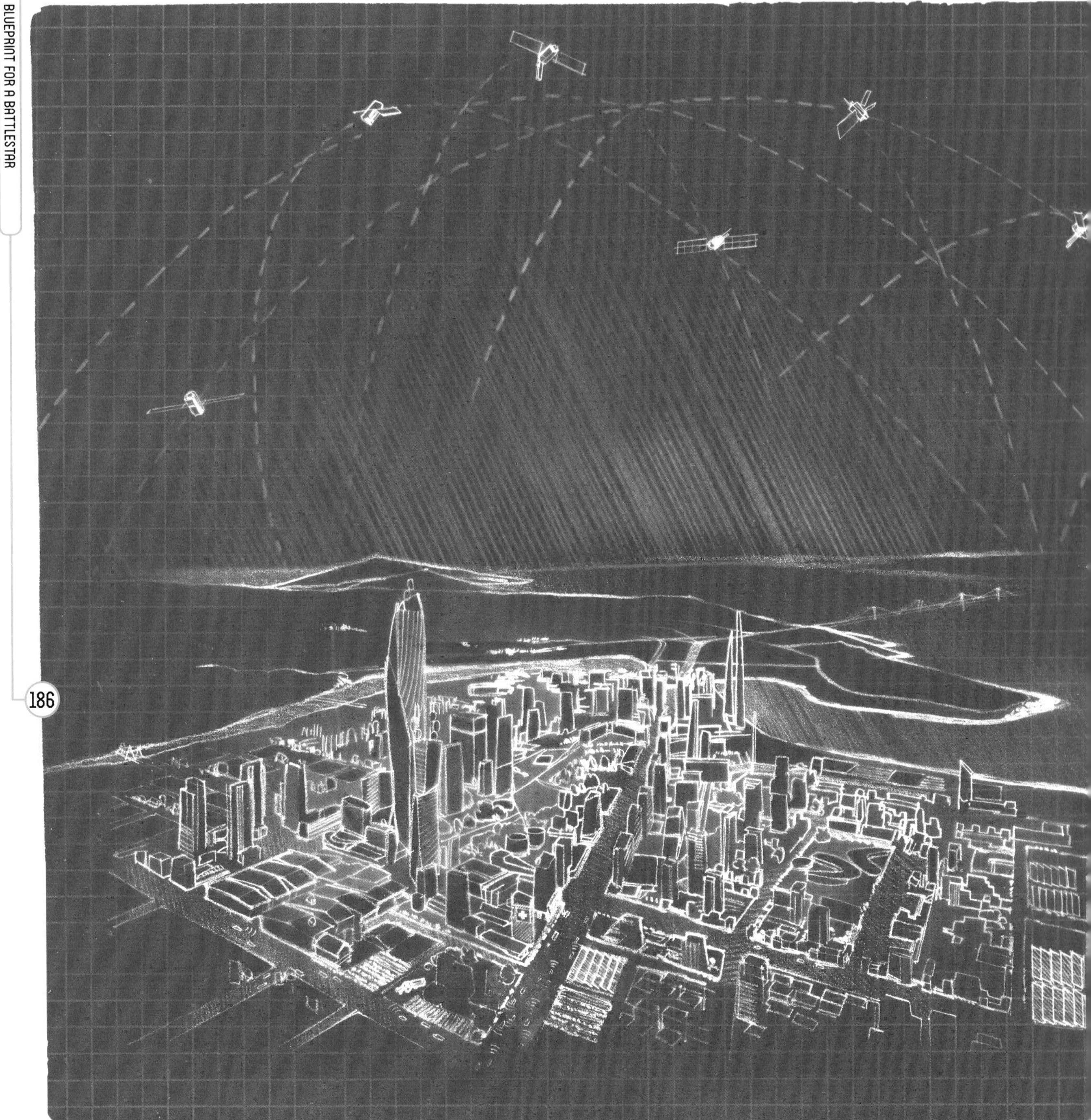

and neurological intervention should minimize the incidence of most crimes. As our database on human behavior and societal surveillance continues to expand, and data mining and interpretation improves, much crime should be predictable. We may end up living in a world where potential perpatrators are identified prior to the committing of a crime and rehabilitated before they have the opportunity to transgress, which will bring with it a new set of ethical challenges.

In the cases that individuals cannot be 'cured' of their violent or predatory nature, society will reach yet another divergence of choice: true rehabilitation (via whatever technologies are available,) continued incarceration, or even elimination. Let's hope that workable solutions can be found for the conditions

WIRED

The communities of the future will be increasingly dependent upon interconnectedness to optimize convenience and lessen environmental impact. Many of our in-home devices—kitchen appliances, lighting, entertainment systems, heat and air conditioning, and more will be continuously monitored and connected to our smartphones, commercial service providers, and security services. AI 'butlers' will become commonplace. Critical to this goal is increased capability for communication traffic, which will be accomplished via hundreds of new and inexpensive 'micro-satellites' and high-flying telecommunication drones that permanently circles the skies above populated areas.

—physical, social, and mental—that cause people to turn to crime.

Nations will maintain their militaries, but there will be few with a truly global reach. Those remaining mega-militaries will be so technology-rich that what altercations do flare up will be dealt with quickly. And it's a good thing, because war will become so expensive that only a few nations will be able to afford the vast investments needed to engage in combat. When they do, it will have to be brief and decisive, or it will rupture their budgets. And woe to those who engage in terrorism; the response will be so individual and instantaneous as to be truly terrifying. Various kinds of fundamentalism may continue to be a driver of small-scale conflict, but there will be few places for perpetrators to hide due to complete, global surveillance and information sharing. It's a future that has the potential to be both frightening and beneficent...possibly both.

This is a reasoned, though optimistic, vision of tomorrow.Many don't want to believe that the future can be full of promise and plenty. They argue that human nature will continue to drive people in power to crave ever more influence and riches, and that any society needs a permanent underclass. Some think that most folks just want handouts, that people are lazy and unmotivated, that humans will always fall prey their baser instincts, and that the predators will always rule the pack, taking from those weaker than themselves. They think that technology will increasingly be used by the governments and militaries of the world to remove our privacy and freedoms, and to bend us to the will of an evermore powerful and despotic few who already control most of the world economy. And they may be right, because no matter how advanced the technology and how amazing the potential, the promise of humanity is that of choice, to use for the general welfare or for the enrichment of the few.

In the end, it's really up to us how this future rolls out. We, as a society, will have to decide what is ethical and what is not, what is safe and what is not, and, perhaps most importantly, who is capable of leading us into a safe, prosperous, and rewarding technological future. And that may end up being the largest challenge of all.

Of course, we could just wait a decade or two and leave it up to our artificial intelligence-enhanced, cybernetic overlords, but I say, let's take responsibility for ourselves and make some hard choices before we get there. It's really up to us, after all.

ACKNOWLEDGMENTS AND PICTURE CREDITS

Acknowledgements are due to the capable Casey Handmer, who recently completed his PhD in theoretical astrophysics at Caltech and came to me by way of the wonderful Michele Judd at the Keck Institute for Space Studies (KISS). Casey fact checked each chapter and kept me honest (mostly). Speaking of KISS, if you want to see what the future of spaceflight holds, it's a great place to check out (http://kiss.caltech.edu/).

The assorted editors, illustrators, and others at Aurum Books created the project and made it happen—kudos especially to Jennifer Barr, Senior Commissioning Editor there. John Willig, agent extraordinaire, was helpful throughout as always.

Sherry Clark provided daily inspiration. Susan Holden Martin was there to provide a boost when needed. Mom, dad, Gloria, Connor, Cathie, Ken, Scott...you're all part of the endeavor too. But this is beginning to sound like the last scene in "The Wizard of Oz" (" You, and you, and you were there!") so I'll stop now.

Publisher: Richard Green

Commissioning Editor: Jennifer Barr

Design: Drew McGovern, Punch Bowl Design

Production Controller: John Casey

Additional Editorial Work: Daniela Rogers, Caroline Curtis, Philip Parker.

Illustration Credits:

Picture Credits:

Every effort has been made to credit the copyright holders of the images used in this book. We apologise for any unintentional omissions or errors and would be pleased to insert the appropriate acknowledgement to any companies or individuals in any subsequent editions of the work.

Alamy (Andrey Volodin) 114, (I. Glory) 78, 100, (RGB Ventures/ SuperStock) 42-3, (Stocktrek Images, Inc.) 14, (The Natural History Museum) 122; Collection of George Dyson, image via https://www.flickr.com/photos/xeni/272465095/in/photostream. For more information see: http://boingboing.net/2006/10/17/project_orion_more_c.html 60; DARPA 105; ESA/ATG Medialab 24-5; Getty Images (a-r-t-i-s-t) 166, (Barry Downard) 128, (Erik Simonsen) 16, 22-3, (gremlin) 182, (Hoberman Collection/UIG) 123, (Sebnem Coskun/Anadolu Agency) 74, (STAN HONDA/AFP) 120, (Tetra Images) 36-7, (ullstein bild) 26, (VICTOR HABBICK VISIONS) 68, 73, 84, (Witold Skrypczak) 158-9, (Zap Art) 152-3; HEADO 134; NASA (JPL/MSSS) 90; Sandia National Laboratories 109; SCIENCE PHOTO LIBRARY (Amelie-Benoist / BSIP) 112, (BRIAN BELL) 140, (CHRISTIAN DARKIN) 116, (DETLEV VAN RAVENSWAAY) 142, 150-1, (HENNING DALHOFF) 52, (MPI BIOCHEMISTRY/ VOLKER STEGER) 119, (SCIEPRO) 106, (WALTER MYERS) 96-7; Shutterstock.com (mik ulyannikov) 32-3, 58; U.S. Department of Energy 174.

FURTHER READING

Kaku, Michio: *Physics of the Future: How Science Will Shape Human Destiny and Our Daily Lives by the Year 2100*. Knopf Doubleday, NY, 2011.

Kurzweil, Ray: *The Singularity is Near: When Humans Transcend Biology*. Penguin Books, NY. 2005.

K. Eric Drexler: *Radical Abundance: How a Revolution in Nanotechnology Will Change Civilization*. Hachette Book Group, NY. 2013.

Shostak, Seth: *Confessions of an Alien Hunter: A Scientist's Search for Extraterrestrial Intelligence*. National Geographic, DC. 2009.

Horner, Jack and Gorman, James: *How to Build a Dinosaur: The New Science of Reverse Evolution*. Plume Publishing, NY. 2010.

Aldrin, Buzz and David, Leonard: *Mission to Mars: My Vision for Space Exploration*. National Geographic, NY. 2015.

Negroponte, Nicholas: *Being Digital*. Vintage Books, NY. 1995.

Krauss, Lawrence: *The Physics of Star Trek*. Basic Books, NY. 2007.

SOURCES

Chapter 1 Death From Above

'A NASA Engineer Explains How to Build a Death Star.' Rob Bricken, *WIRED*, December 15, 2015. http://www.wired.com/2015/12/nasa-death-star-asteroid/

'The Death Star: Could It Destroy a Planet?' Jeanne Cavelos, *Scientific American*, August 2008.

'The White House's nerd-delighting Death Star petition response.' Chris Gayomali, *The Week*, January 14, 2013.

'Military Technology: Laser weapons get real.' Andy Extance, *NATURE*, May 2015.

Military Laser Systems. Jon Grossman, RAND Corporation, 1991.

Building Large Structures in Space. Tennessee Valley Interstellar Workshop presentation, November, 2014.

Chapter 2 The Ultimate Weapon

Nato Handbook On The Medical Aspects Of Nbc Defensive Operations. Army Field Manual 8-9, Navy Medical Publication 5059, Air Force Joint Manual 44-151. 1996.

'Microwave weapons: Wasted.' Sharon Weinberger, *Nature*, September 12, 2012.

'High-energy Laser Weapon Burns Through a Running Engine From a Mile Away.' *Defense Update*, March 6, 2015. http://defense-update.com/20150306_laser_weapon.html#.VjkmeK6rSRs

'High Energy Laser Directed Energy Weapons, Technical Report APA-TR-2008-0501.' Carlo Kopp, Air Power Australia, May 2008. http://www.ausairpower.net/APA-DEW-HEL-Analysis.html

'Boeing's Hel Md.' Jack Satterfield, *Tactical Life*, March 1, 2013. http://Www.Tactical-Life.Com/Firearms/Boeings-Hel-Md/

'Tesla on High Frequency Generators.' Electrical Experimenter, April 1919.

'Development of the Chemical Oxygen-Iodine Laser (COIL) with chemical generation of atomoic iodine.' J. Kodymova, O. Spalek, V. Jirasek, M. Censky, G.D. Hager, *Appiled Physics*, 2003.

Space Warfare and Defense. Bert Chapman, ABC Clio, 2008.

Chapter 3 Shields Up!

'Plasma bubble could protect astronauts on Mars trip.' David Shiga, *New Scientist*, July 17, 2006.

'Boeing to Develop Plasma Force Field. Military Connection,' April 12, 2015. http://www.militaryconnection.com/blog/5303/military-connection-boeing-develop-plasma-force-field/

'*Star Trek*-style force-field armour being developed by military scientists.' Richard Gray, the *Telegraph*, March 20, 2010.

Chapter 4 Swords of Heat

'How a Plasma Cutter Works.' Techmate, Torchmate, 2015. http://www.torchmate.com/

'Humming, Glowing, Glorious Lightsabers.' Adrienne LaFrance, *The Atlantic*, December 2015.

'Are Lightsabers Possible?' Physics.org. http://www.physics.org/article-questions.asp?id=59

'Seeing Light in a New Way.' Peter Ruell, *Harvard Gazette*, September 2013.

'Holographic Ruby Laser with Long Coherence and Precise Timing.' M. Young and A. Hicks, *The Opical Society*, vol.13, issue 11, 1974.

Chapter 5 Judgment Day

'Stephen Hawking, Elon Musk, and Bill Gates Warn About Artificial Intelligence.' Michael Sainato, the *Observer*, August 19, 2015.

'Top notch AI system about as smart as a four-year-old, lacks commonsense.' Darren Quick, *Gizmag*, July 15, 2013. http://www.gizmag.com/ai-system-iq-four-year-old/28321/?li_source=LI&li_medium=default-widget

'Google's Articifical Intelligence can Probably Beat You at Video Games.' Tanya Lewis, LiveScience, February 25, 2015.http://www.livescience.com/49947-google-ai-plays-videogames.html

Chapter 6 Stun, Kill, or Disintegrate

'Real-Life Laser Rifle: Army Goal.' David Hambling, *WIRED*, April 27, 2007.

'US military sets laser PHASRs to stun.' Will Knight, *New Scientist*, November 2005.

'The Army Is Testing Handheld Ray Guns.' Patrick Tucker, DefenseOne, April 2015. http://www.defenseone.com/technology/2015/04/army-testing-handheld-ray-guns/110815/

Beam Weapons: The Next Arms Race. Jeff Hecht, Jeff Hecht. 1984.

Beam: The Race to Make the Laser: Jeff Hecht, Oxford University Press, 2005.

A Practical Introduction to Homeland Security and Emergency Management. Bruce Oliver Newsome, Jack Jarmon, Sage Publications, 2016.

Chapter 7 Rocket Science

Bell Rocket Belt. Skytamer Archive, 2016. http://www.SKYTAMER.COM/BELL_ROCKET_BELT.html

'The Ill-Fated History of the Jet Pack.' Jeff McGregor, *Smithsonian*, June 2015.

'Up, up and down: The ephemerality and reality of the jetpack.' John Turi, *Smithsonian*, June 2015

'Here's Your Jetpack.' Chris Higgins, Mentalfloss, September 17, 2012. http://mentalfloss.com/article/12559/heres-your-jetpack

Master Mechanics and Wicked Wizards. Glenn Scott Allen, University of Massachusetts Press, 2009.

Chapter 8 Wormholes

'Is Warp Drive Real?' March 9, 2015, Glenn Research Center, NASA. http://www.nasa.gov/centers/glenn/technology/warp/warp.html

'Starship Concepts.' Tau Zero Foundation, 2016. https://tauzero.aero/home/

'It's Time to Tackle Interstellar Spaceflight, Experts Say.' Mike Wall, Space.com, June 5, 2013. http://www.space.com/21452-interstellar-spaceflight-starship-congress.html

Warp Speed? Not So Fast. Matthew R. Francis, Slate, August 23, 2013.

http://www.slate.com/articles/health_and_science/science/2013/08/harold_sonny_white_warp_drive_faster_than_light_secret_physics_debunked.html

'Interstellar Spaceflight Using Nuclear Propulsion and Advanced Techniques.' Sateesh Pande, Ugar Guven, Gurunadh Velidi, 62nd International Astronautical Congress, Cape Town, South Africa, 2011.

Chapter 9 Space-Aged Wagon Trains

'Starfleet was closer than you think.' Brent Ziarnick and Lt. Col. Peter Garretson, *Space Review*, March 16, 2015.

'Interstellar Flight Visionaries Gather for the 100 Year Starship Symposium.' Charles Black, SEN, September 2012.http://sen.com/news/100-year-starship-symposium-considers-interstellar-travel

'There and Back Again.' Henry Garrett, Keck Institute for Space Studies, July 2012. http://www.kiss.caltech.edu/workshops/systems2012/presentations/garrett.pdf

Chapter 10 They Don't Work on Water

'How the Most Promising Hoverboards Actually Work.' Rhett Allan, *WIRED*, October 20, 2015.

'Magnetic levitation of a stationary or moving object.' Patent, US Patent and Trademark Office, 2014.

'Engineering the Future.' Arcaspace, 2015. http://www.arcaspace.com/

Superconductivity. Charles Poole Jr, Horacio Farach, Richard Creswick, Ruslan Prozorov, Academic Press, 2007.

Chapter 11 Are We There Yet?

'The Flying Car That Could Expedite Your Morning Commute.' Carl Engelking, *Discover Magazine*, May 1, 2014. http://discovermagazine.com/authors?name=Carl+Engelking

'Paul Moller: My Dream of a Flying Car.' Paul Moller, Ted Talk, February 2004. http://www.darpa.mil/program/aerial-reconfigurable-embedded-system

Aerial Reconfigurable Embedded System (ARES). Ashish Bagai, DARPA. http://www.darpa.mil/program/aerial-reconfigurable-embedded-system

'Flying Cars.' Carnagie Mellon University, University News, 2013. http://www.cmu.edu/homepage/computing/2010/fall/flying-cars.shtml

'Flying Car.' Jeremy Hsu, *Scientific American*, 2009. http://www.nature.com/scientificamerican/journal/v310/n4/full/scientificamerican0414-28a.html

Chapter 12 Living in Space

'Settling space is the only sustainable reason for humans to be in space.' Dale Skran, *The Space Review*, February 1, 2016 http://www.thespacereview.com/article/2915/1

'Using space resources to help all of humanity.' Greg Anderson, *The Space Review*, February 1, 2016. http://www.thespacereview.com/article/2913/1

The High Frontier: Human Colonies in Space. Gerard K. O'Neill and Freeman Dyson., Apogee Press, 2000.

'Made in space: new production frontiers.' Andrew Wade, *The Engineer*, February 2016.https://www.theengineer.co.uk/made-in-space-new-production-frontiers/

'Orbital Space Settlements.' The National Space Society, 2014. http://www.nss.org/settlement/space/

Space Settlement Population Rotation Tolerance. Al Globus, San Jose State University, and Theodore Hall University of Michigan, 2015.

Chapter 14 More than Human

'Bionic Skin for a Cyborg You: Flexible electronics allow us to cover robots and humans with stretchy sensors.' Takao Someya, IEEE SPECTRUM Online, August 2013. http://spectrum.ieee.org/biomedical/bionics/bionic-skin-for-a-cyborg-you

'DARPA heads for robot-human hybrid: Are cyborgs on the way?' James Maynard. *Tech Times*, April 3, 2014. http://www.techtimes.com/articles/5137/20140403/darpa-robot-human-hybrid-cyborgs.htm#sthash.xaE2SPpS.dpuf

'Could Human Enhancement Turn Soldiers Into Weapons That Violate International Law? Yes.' Patrick Lin, *The Atlantic*, January 4, 2013.

'Kansei Engineering and Soft Computing: Theory and Practice.' Ying Dai, *Engineering Science Reference*, Hershey, PA.2011.

'Modest Debut of Atlas May Foreshadow Age of "Robo Sapiens"' John Markoff, the *New York Times*, July 11, 2013. http://www.nytimes.com/2013/07/12/science/modest-debut-of-atlas-may-foreshadow-age-of-robo-sapiens.html?_r=1

Chapter 15 Atomic Manufacturing

'Nanocarriers may carry new hope for brain cancer therapy.' Ting Xu, University Of California At Berkeley, November 2015.http://www.nanotech-now.com/news.cgi?story_id=52611

'The incredible "flowering" nanomachines that can bend themselves into shape. Mark Prigg,' the *Daily Mail*, October 19, 2012. http://www.dailymail.co.uk/sciencetech/article-2220282/The-astonishing-flowering-micromachines-build-themselves.html

Regulating Technologies: Legal Futures, Regulatory Frames and Technological Fixes. Roger Brownsord (editor) and Karen Yeung (editor), Hart Publishing, 2008.

Chapter 16 Man or Machine?

'Tiny, Logical Robots Injected Into Cockroaches.' Jesse Emspak, LiveScience, April 9, 2014. http://www.livescience.com/44704-nanobots-injected-into-cockroaches.html#sthash.k9pvwX7r.dpuf

'Military's "Iron Man" Suit Debuts This Month.' Tanya Lewis, LiveScience, May 27, 2014. http://www.livescience.com/45860-ironman-suit-military-tech.html

Beyond Human: Living with Robots and Cyborgs. Gregory Benford, Forge Books, 2008.

Natural-born Cyborgs: Minds, Technologies, and the Future of Human Intelligence. Andy Clark, Oxford University Press, 2003.

Creative Creatures: Values and Ethical Issues in Theology, Science and Technology. Ulf Gorman (editor), William Drees, Hubert

Meisenger, Bloomsbury T&T Clark, 2005.

Chapter 17 My Pet T-Rex

'Reverse engineering dinosaurs.' Zartasha Mustansar, Lee Margets, Phillip Manning, Hillel Kugler, University of Manchester Conference presentation (via Researchgate,) October 2015.

'Reverse Engineering Birds' Beaks Into Dinosaur Bones' by Carl Zimmer, the *New York Times*, May 12, 2015.

The Montana Professor. Jack Horner, Montana State University, April 9,2011.

How to build a dinosaur: extinction doesn't have to be forever. John Horner and James Gorman, Dutton Books, 2009.

'Researchers Created a DinoSkulled Chicken to Explore Evolution.' R.A. Becker, PBS.org, May 2015. http://www.pbs.org/wgbh/nova/next/evolution/scientists-created-a-dino-skulled-chicken-to-explore-evolution/

Chapter 18 Hands-Free

'USC Lab Creates 3-D Holographic Displays, Brings TIE Fighters to Life.' Jose Fermoso, *WIRED*, June 26, 2008.

'Neurological evidence linguistic processes precede perceptual simulation in conceptual processing.' Max Louwerse and Sterling Hutchinson, *Frontiers in Psychology*, October 2012. http://journal.frontiersin.org/article/10.3389/fpsyg.2012.00385/full

'Exclusive: Google's New Head of Virtual Reality on What They're Planning Next.' *TIME Magazine*, May 26, 2016.

'Finite Element Model for femtosecond laser pulse heating using dual-phase lag effect.' S. Kumar, S. Bag, M. Burah, *Journal of Laser Applications*, 2016.

'Gesture in Human-Computer Interaction and Simulation.' Sylvie Gibet, Nicolas Courty, Jean-Francois Kamp, 6th International Gesture Workshop, Springer Berlin Heidelberg, Sep 2, 2009

Face Detection and Gesture Recognition for Human-Computer Interaction. Ming-Hsuan Yang, Narendra Ahuja, Springer Books, 2001.

The Human-Computer Interaction Handbook. Andrew Sears (editor,) Julie A. Jacko (editor), Lawrence Erlbaum Associates/CRC Press, 2006.

Chapter 19 Auto-Diagnosis

'Tricorder Xprize.' Xprize Foundation, 2016. http://tricorder.xprize.org/

'Scanadu Scout: A real-life "tricorder" brings a touch of "*Star Trek*" to medicine. Martyn Williams, *PC Magazine*, August 12, 2015.

'X-Rays and Tricorders; Futuristic Hand-held Medical Diagnostics.' Kyle Maxey, Engineering.com, January 11, 2013. http://www.engineering.com/DesignerEdge DesignerEdgeArticlesArticleID/5162/X-Rays-and-Tricorders-Futuristic-Hand-held-Medical-Diagnostics.aspx

The Tricorder Project. http://www.tricorderproject.org.

'X-Rays and Tricorders; Futuristic Hand-held Medical Diagnostics.' Kyle Maxey, Engineering.com, January 11, 2013. http://www.engineering.com/DesignerEdge DesignerEdgeArticles/ArticleID/5162/X-Rays-and-Tricorders-Futuristic-Hand-held-Medical-Diagnostics.aspx

Chapter 20 Eco-Engineering

'Making Mars the New Earth.' *National Geographic Magazine*, January 2010.

Planetary Ecosynthesis on Mars: Restoration Ecology and Environmental Ethics. Christopher P. McKay, NASA Ames Research Center, December 2007.

Technological Requirements for Terraforming Mars. Christopher P. McCay and Robert Zubrin, NASA/STI, 2007.

Terraforming. Martyn Fogg, Society of Automotive Engineers, 1995.

The Case for Mars. Robert Zubrin, Simon & Schuster, 2011.

Chapter 21 Plugging In

'A Brain-Computer Interface that Works Wirelessly.' Antonio Regalado, *MIT Technology Review*, January 2015. https://www.technologyreview.com/s/534206/a-brain-computer-interface-that-works-wirelessly/

'The Next Generation In Human Computer Interface.' Singularity Hug, Singularity University, March 4, 2014. http://singularityhub.com/2009/03/04/the-next-generation-in-human-computer-interfaces-awesome-videos/

'Scientists believe they have come close to solving the "Matrix" theory.' Lucy Kinder, the *Telegraph*, October 26, 2012.http://www.telegraph.co.uk/news/science/9635166/Scientists-believe-they-have-come-close-to-solving-the-Matrix-theory.html

Chapter 22 Who's Out There?

'The Optimistic Gamble.' *The Economist*, July 25, 2015. http://www.economist.com/news/science-and-technology/21659487-bold-new-programme-financed-silicon-valley-tycoon-will-revitalise-hunt

'The History of SETI.' SETI.ORG, 2016. http://www.seti.org/node/662

'In Search of ET.' Jane Hu, *Berkeley Science Review*, June 2014. http://berkeleysciencereview.com/article/search-e-t/

'SETI: peering into the future.' Alan Penny, Oxford Journals, 2011.http://astrogeo.oxfordjournals.org/content/52/1/1.21.full

Chapter 23 Stopped Cold

'Telomere extension turns back aging clock in cultured human cells.' Helen Blau, *Stanford Medicine News*, January 2015. https://med.stanford.edu/news/all-news/2015/01/telomere-extension-turns-back-aging-clock-in-cultured-cells.html

'Are Telomeres The Key To Aging And Cancer? Learn Genetics,' University of Utah. http://learn.genetics.utah.edu/content/chromosomes/telomeres/

'Telomerase as an emerging target to fight cancer Opportunities and challenges for nanomedicine'. Lorenz Philippi, UFSchaefer, CM Lehr; Loretz; Schaefer; Lehr, Journal of Controlled Releases, 2010.

'Comparative biology of mammalian telomeres: hypotheses on ancestral states and the roles of telomeres in longevity determination'. Gomes, NM; Ryder, OA; Houck, ML; Charter, SJ; Walker, W; Forsyth, NR; Austad, SN; Venditti, C; Pagel, M; Shay, JW; Wright, WE. 2010.

Chapter 24 Disappearing Act

'Berkeley Lab Team Develops Ultrathin, Flexible Gold-Nano-Brick "Invisibilty Cloak": Adaptive Camouflage/Visual Cloaking Tech for any 3D Object.' David Crane, *Defense Review*, September 2015. http://www.defensereview.com/berkeley-lab-team-develops-ultrathin-flexible-gold-nano-brick-invisibilty-cloak-adaptive-camouflagevisual-cloaking-tech-for-any-3d-object/

'Ultrathin "Invisibility Cloak" Can Match Any Background.' Jesse Emspak, LiveScience, September 17, 2015. http://www.livescience.com/52216-ultrathin-invisibility-cloak.html

'The US Army Is Serious About Developing Invisibility Cloaks.' Akshat Rathi, DefenseOne, May 8, 2015. http://www.defenseone.com/technology/2015/05/us-army-serious-about-developing-invisibility-cloaks/112291/

'Invisibility System Using Image Processing and Optical Camouflage Technology.' Vasireddy Srikanth , Pillem Ramesh, *International Journal of Engineering Trends and Technology*, Volume 4 Issue 5, May2013

Chapter 25 Building Tomorrow

'Introduction to Arcology.' Arcosanti.org, 2016. https://arcosanti.org/Arcology

Sentient City: Ubiquitous Computing, Architecture, and the Future of Urban Space. Mark Shepard, MIT Press, 2011.

Cities of the Future. Jacob Morgan, Forbes, September 4, 2014.

'Information Arcology and Data Explorations: Scientific Content for Multiple Learning Styles and Environments.' Ted Habermann, Nancy Burton, Kevin Frender, *Journal of Scientific and Education Technology*, Volume 7, Issue 3, September 1998.

'High-Rise Complex Named Arcology Skyscraper Designed for Hong Kong Harbor.' Council on Tall Buildings and Urban Habitat, March 2016.

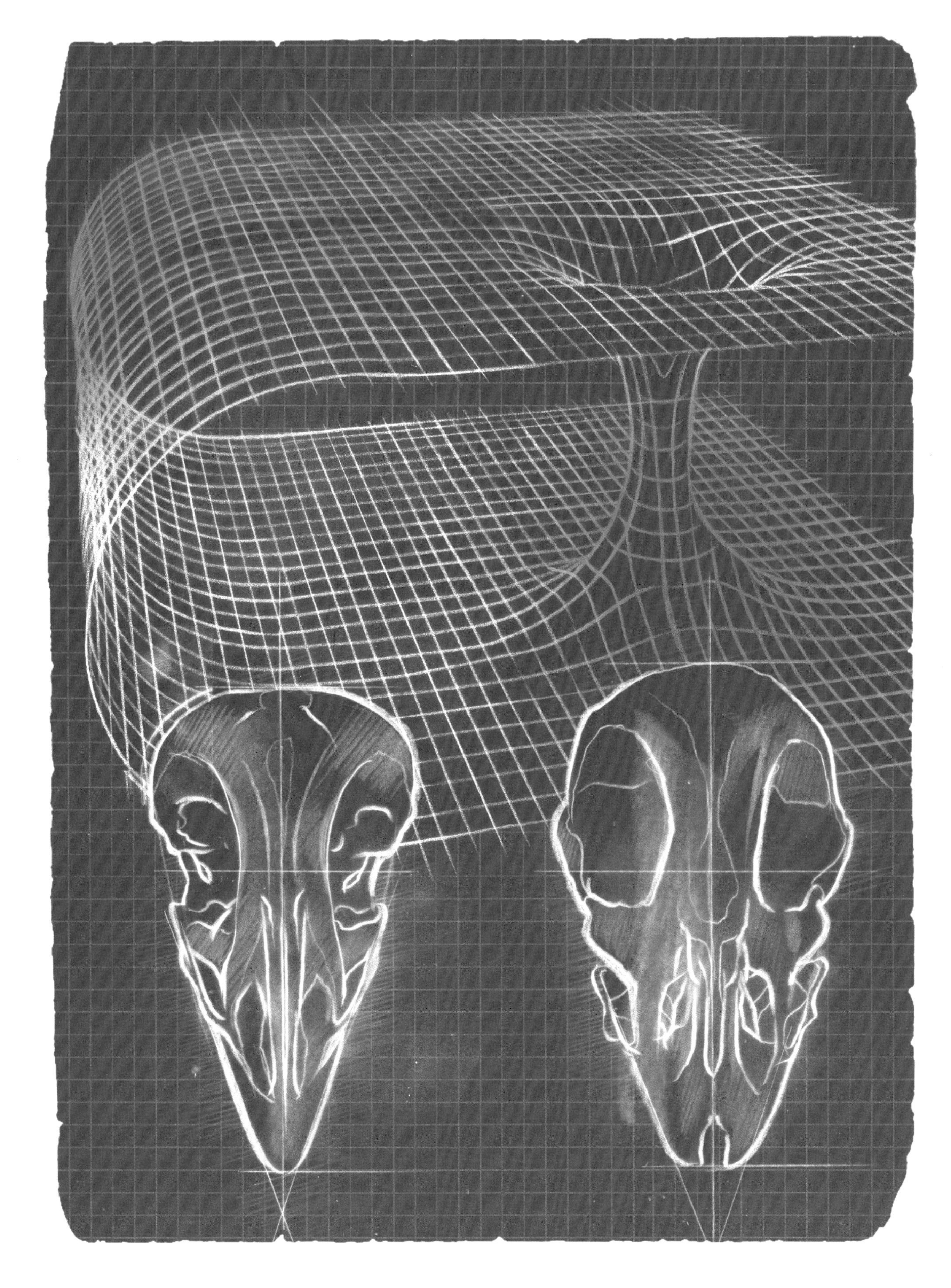